DANCE FOR TWO

DANCE

FOR TWO

Selected Essays

ALAN LIGHTMAN

PANTHEON BOOKS

NEW YORK

All rights reserved under International and Pan-American Copyright Conventions.
Published in the United States by Pantheon Books, a division of Random House,
Inc., New York, and simultaneously in Canada by Random House of
Canada Limited, Toronto.

"Progress" was originally published in *Inc. Magazine*.

All other essays were originally published in the following books: *Time Travel and
Papa Joe's Pipe*, copyright © 1984 by Alan P. Lightman (New York: Charles
Scribner's Sons) and *A Modern Day Yankee in a Connecticut Court*, copyright © 1984,
1985, 1986 by Alan Lightman (New York: Viking Penguin Inc.).

Library of Congress Cataloging-in-Publication Data
Lightman, Alan P., 1948–
Dance for two ; essays / Alan Lightman.
p.   cm.
ISBN 0-679-75877-1
1. Science—Popular works.   I. Title.
Q162.L53   1996
500—dc 20   95-48876
CIP

Book design by Cathryn Aison
Manufactured in the United States of America
First Edition
2 4 6 8 9 7 5 3 1

To Jean, Elyse, and Kara

Contents

Contents

FOREWORD

THE TWENTY-FOUR ESSAYS in this collection were chosen from the last fifteen years; all have had a previous life in magazines and in other collections. These are the pieces that have made me happiest in the rereading and seem the most likely to keep doing so. Over time, I have come to understand there are three pleasures in writing. The first is the exquisite joy of writing itself, when one is completely alone; the second, more social, comes in moving readers with what one has written; the third pleasure, years later and finally solitary again, is rereading that fraction of one's writing worth keeping and being surprised and grateful all over again. For the most part, writing is a selfish and self-centered profession. And the essay, as E. B. White has remarked, is probably the most egoistic of all forms. Here, the writer openly displays personal thoughts and adventures, as if his every sneeze and little observation will be of general interest.

The idea for my first essay came to me while I sat in a comfortable wingback chair, smoking my great-grandfather's pipe and inhaling the ancient aromas long dormant within it. As described in "Time Travel and Papa Joe's Pipe," the pipe created a kind of intimacy with my ancestor, deceased before I was born, and set my mind puzzling about time travel. More importantly, it revived my relationship with my father. He was a pipe smoker himself and a taciturn man. For years, I had never known what my father was thinking, whether he was pleased or unhappy. But beginning in my college years, he had now and then presented me with one of his pipes and a brief story to go with it. Once he had given me a light-brown Kaywoodie from his World War II days and described how he used to smoke it while pacing on his ship before an invasion. My great-grandfather's old briar, with its strange engraved markings, had rested in his drawer for years before he gave it to me, without comment. Then, a few days later, he sent me a wonderful photograph of himself as a little boy with Papa Joe, just the two of them, holding hands in front of a white clapboard house. My father wore knickers and Papa Joe a hat and mustache, just as I'd pictured him from the aromas in his pipe. I wrote my essay and mailed it to my father. Then, miraculously, we began truly talking to each other. And I, well on the way to becoming as quiet as my father, discovered that through writing I could open up myself and touch people I cared about.

That initial essay (as well as a second) was published in *Smithsonian* magazine. Then I began a long-running monthly column for the excellent but now defunct *Science 80*; then I wrote for other magazines when *Science 80* went out of business in the mid-1980s. Encounters with magazines sober a beginning writer, who is often so enamored of his scribblings that he's committed every word to memory. In magazines, sentences and whole paragraphs are often slashed by an editor to make way for the next story or even a cartoon embedded in the page for distraction. Despite this abuse, the writer continues to write, the habit being so insatiable and compulsive.

From the beginning, I found myself writing of science, my first passion and profession, sometimes of the hard facts of science but more often of the human texture and whimsy, the lived part of science. Science, for me, was the most rigorous and extreme expression of order in the physical world. Yet the desire for that order, and often the means to declare it, were human, oddly nestled against the emotion and wild flight of the human world. Where those two worlds met seemed a subject for literature. And I was partly propelled by something I'd learned from watching my colleagues: Scientists often make their greatest discoveries just at those moments when they follow their intuition instead of equations. In other words, when they behave the least "scientifically." That secret, known to historians but rarely to scientists, became the hidden thread running through my essays.

While writing, I became fascinated by the creative tension between science and art, reason and instinct. Suspecting a great deal of instinct in science and reason in art, I asked science friends if they ruminated in pictures or equations, to what extent did they use aesthetic criteria in their work, if they believed in metaphors. I asked artist friends how ideas came to them, how paintings were balanced, why a particular splash of color was placed where it was. I came back to Einstein's provocative comment that there is no logical path to the laws of nature, that the laws can be reached only by intuition and the "free inventions of the mind." Could a scientist invent the world like an artist? Wasn't there still a world outside of our minds? People had landed on the moon and returned. Which world was true?

One day, during the midst of my questions, I took my two-year-old daughter to the ocean for the first time. It was a mild, slightly hazy day in June. We parked our car a half mile from the water and walked toward the coast. A speckled pink crab shell lay on the sand and caught her attention. A hundred yards farther, we heard the rolling of the waves, in rhythmical sequence, and I could tell that my daughter was curious about what made the sound. Holding her up with one arm, I pointed to the sea. My daughter's eyes followed along my arm, across the beach, and then out to the vast blue-green ocean. For a moment she hesitated. I wasn't sure whether she would be puzzled or frightened by that first sight of infinity. Then she broke out into a radiant smile. There was nothing I needed to say to her, nothing I needed to explain.

The essay is well suited to my own unsettled identity as a scientist and writer. It is a generous form of writing, able to accommodate the philosopher, the teacher, the polemicist, the raconteur, the poet. All one requires is an initial idea, the willingness to become personal about the subject (often one's self), and the discipline to shut up before the composition becomes a book. The subject of science poses special challenges to the essayist, for most people want to read about people, or at least things tied to people. Much of science, of course, is inanimate and far removed from daily life. In this regard, an essay about medicine or psychology may be intrinsically more engaging than one about chemistry or physics. People must be put into writing on science the same way M.F.K. Fisher puts them into her writing about food: What is the ideal number of people to invite to a dinner party, and why? What about the funny mustache of the waiter who served Ms. Fisher in that little restaurant in Dordogne. By the time she gets to the food, which invites on its own, our appetite has been whetted.

In many of these essays, science is merely a jumping off point into the uncertain terrain of human behavior. Nearly half of the pieces are part parable or fable or story. Beyond its particular subject matter, every piece requires a fresh approach, as much to entertain the writer as the reader. Like a short story, the conception of an essay either works or it doesn't, and if it doesn't one must throw the thing into the wastebasket without pity. I hope I have thrown away all that I should.

DANCE FOR TWO

Pas de Deux

In soft blue light, the ballerina glides across the stage and takes to the air, her toes touching Earth imperceptibly. *Sauté, batterie, sauté.* Legs cross and flutter, arms unfold into an open arch. The ballerina knows that the easiest way to ruin a good performance is to think too much about what her body is doing. Better to trust in the years of daily exercises, the muscles' own understanding of force and balance.

While she dances, Nature is playing its own part, flawlessly and with absolute reliability. On *pointe*, the ballerina's weight is precisely balanced by the push of floor against shoe, the molecules in contact squeezed just the right amount to counter force with equal force. Gravity balanced with electricity.

An invisible line runs from the center of the Earth through the ballerina's point of contact and upward. If her own center should drift a centimeter from this line, gravitational torques will topple her. She knows nothing of mechanics, but she can hover on her toes for minutes at a time, and her body is continuously making the tiny corrections that reveal an intimacy with torque and inertia.

Gravity has the elegant property of accelerating everything equally. As a result, astronauts become weightless, orbiting Earth on exactly the same trajectories as their spaceships and thus seeming to float within. Einstein understood this better than anyone and described gravity with a theory more geometry than physics, more curves than forces. The ballerina, leaping upward lightly, hangs weightless for a moment amid flowers she has dropped midair, all falling on the same trajectory.

Now she prepares for a *pirouette*, right leg moving back to fourth position, pushing off one foot, arms coming in to speed the turn. Before losing balance she gets four rotations. Male dancers, on *demi-pointe* and with greater contact area, can sometimes go six or eight. The ballerina recovers well, giving her spin smoothly back to Earth and remembering to land in fifth position smiling. Briefly her feet come to rest, caught between the passage of spin and the friction of the floor. Friction is important. Every body persists in its state of rest or of uniform motion unless acted upon by outside forces. Every action requires a reaction.

The ballerina depends on the constancy of the laws of physics, even though she herself is slightly unpredictable. In this same performance last night she went only three and a half turns through her first *pirouette*, and then took the *arabesque* several feet from where she takes it now. Regardless of these discrepancies, the atoms in the floor, wherever she happens to touch and at one millisecond's notice, must be prepared to respond with faithful accuracy. Newton's laws, Coulomb's force, and the charge of electrons must be identical night after night—otherwise, the ballerina will misjudge the resiliency of the floor or the needed moment of inertia. Her art is more beautiful in its uncertainty. Nature's art comes in its certainty.

The ballerina assumes one pose after another, each fragile and symmetrical. In the physics of solids, crystal structures can be found that appear identical after rotations by one-half, one-third, one-quarter, and one-sixth of a circle. Crystals with one-fifth and one-seventh symmetries do not exist because space cannot be filled with touching pentagons or septagons. The ballerina reflects a series of natural forms. She is first ethereal, then lyrical. She has struggled for years to develop a personal style, embellished with fragments from the great dancers. As she dances, Nature, in the mirror, pursues its own style effortlessly. It is the ultimate in classic technique, unaltered since the universe began.

For an ending, the ballerina does a *demi-plié* and jumps two feet into the air. The Earth, balancing her momentum, responds with its own *sauté* and changes orbit by one ten-trillionth of an atom's width. No one notices, but it is exactly right.

A Flash of Light

My serious interest in physics began in my freshman year of college. In the dining hall that year, one of the upperclassmen smugly announced that, on the strength of mechanics alone, he could predict where to strike a billiard ball so that it would roll with no sliding. I was mightily impressed and decided this was a subject worth looking into.

Although I didn't realize it at the time, scientists generally divide into two camps, theorists and experimentalists. The abstractionists and the tinkerers. Especially in the physical sciences, the distinction can be spotted straight off. It has since been my observation that, in addition to their skills in the lab, the experimentalists (particularly the males) can fix things around the house, know what's happening under the hood of a car, and have a special appeal to the opposite sex. Theorists stick to their own gifts, like engaging themselves for hours with a mostly blank sheet of paper and discussing chess problems at lunch. Sometime in college, either by genes or by accident, a budding scientist starts drifting one way or the other. From then on, things are pretty much settled.

My path was decided in junior year. For some reason, the physics department had gotten into its head that we students should have a practical knowledge of our subject. To this end, an ungraded electronics workshop, to be tackled in the fall of that year, was strongly encouraged. Most of my colleagues leaped at the opportunity. This was particularly true of those scholars shaky in course work, who could be heard muttering such quips as, "This will separate the men from the boys." (My college was all male in those days.) I had an inkling of trouble, but was not one to tuck in my tail. I signed up.

This electronics workshop was considerably different from the routine laboratory exercises attached to most courses. In the latter, you were always measuring something where you knew damn well what the right answer was. One experiment I remember involved determining the speed of light. The equipment consisted of two mirrors, one of them stationary and the other rotating rapidly. Light making a round trip between the two mirrors would be slightly deflected on its return path by the rotation of the moving mirror, and from the amount of deflection you could deduce the speed of light. Of course, you could also look up the speed of light in any number of books. If your own measured value came out shy of the mark, you could nudge the mirrors and try again. With

enough stamina, you eventually got the result you were looking for, at which point everything was carefully recorded, the
experiment was declared a success, and you strode out of the
laboratory in search of other mountains to climb.

But this electronics project was different. Each of us was
provided with a large stock of transistors, capacitors, and so
forth, a description of what the final thing had to *do*, and let
loose. The stated aim of my gadget, as I recall, was to light up
for several seconds when pure tones above middle C were
offered to it but to maintain a state of torpor otherwise. (The
state of torpor I had no problem with.) To help us get started
with the fundamentals, we were given a textbook titled *Basic
Electronics for Scientists*, which I immediately recognized as a
friendly ally, took with me everywhere for months, and pored
over deep into the night, at the expense of my own and my
roommate's sleep.

The next couple of months were miserable. I discovered
that what worked in the book didn't necessarily work on the
lab table, at least under my supervision. In this regard, I lagged
far behind most of my classmates. When they looked at the
wavy line of an oscilloscope, it spoke to them, and they knew
just how to fix up their circuits to get the desired results. I
badly wanted my project to succeed. But I did not possess that
peculiar knack for making things work. I could write poetry,
I could play "Clair de Lune" on the piano, and I loved talking
about ideas. But I could not make things work.

One day that term, through some odd misdirection of the
postal service, I received in my mail slot a catalog for a home
electronics course. Normally, I throw such things out. But to

me, at that time, this catalog seemed like a greeting from providence. I took the thing back to my dormitory room, discreetly, and began reading. The front page said something to the effect that, with no prior training or aptitude, you would in six weeks be designing working circuits, mending broken televisions, and presenting yourself to the electronics industry as a force to be reckoned with. There were a few sample diagrams, some pictures of robust-looking devices, and glowing statements by successful graduates. What caught my eye was the provision that, during the course, you could mail in detailed sketches for an electrical project of your own design and be promptly and accurately informed whether the thing would, in fact, work. This last feature was absolutely guaranteed. Mending televisions didn't much interest me, but the chance of securing a foolproof verdict on my floundering electrical enterprise was nothing to be sneered at.

I enrolled without delay in this mail-order electronics course. The cost was $200, and you had to furnish your own parts, of which I had ample supply. My plan was to send in furtively a series of intermediate designs for my college project until one of them received the seal of approval. I could then prance into the physics lab and assemble the contraption in short order. My colleagues, meanwhile, were reporting to the lab daily, laboriously testing out each little step of their assigned projects. I had tried this method and failed. It was a great relief to me that I could now suffer through all the preliminary defeats in private, without humiliating myself in front of the others.

Eventually one of my designs was certified. I spent the last few days before the project deadline calmly at work in the lab, soldering each part into its approved position. My fellow students watched my miraculous progress with the kind of respect that is never verbalized. We were all equals, and I basked in my satisfaction. However, I never had the courage to put the device through a dry run.

The final judgments of the projects were pronounced on a day in December by a highly competent member of the faculty named Professor Pollock. Pollock was a man of few words, but a fair man. He was partially bald, as I remember, wore thick glasses, and usually held his head lowered below eye level. When something you said or did amused him, he would look up briefly and grin, without making the slightest sound or the tiniest movement of his head. Pollock was someone who could make things work. He had large hands. He had built cyclotrons.

On that day in the lab, the various projects and students stood at attention, like dogs and their masters in a dog show. When it came time to put my pet through its paces, I played a note to it—I forget now whether above or below middle C— and it answered with a single, blinding flash of light followed by the unmistakable odor of an electrical fire. The flash going off seemed like a shotgun blast, and I instinctively ducked for cover. It was amazing that no one was hurt. Pollock stood grinning longer than usual.

The following summer Armstrong and Aldrin walked on the moon. As I sat watching them on television in my family house, I had enormous respect for the parts that had worked to get them there: the rocket propellant, the computers, the space suits. And I was filled with admiration for the people behind all that, people good with their hands. Pollock may have been one of them and, undoubtedly, the handy students in my physics class would work on similar things in the future. But it also occurred to me that we theorists were needed to make sure the moon would be there at the same time the astronauts were. There are abstractionists and there are tinkerers, and I was not unhappy to have discovered my lot.

SMILE

IT IS A SATURDAY IN MARCH. The man wakes up slowly, reaches over and feels the windowpane, and decides it is warm enough to skip his thermal underwear. He yawns and dresses and goes out for his morning jog. When he comes back, he showers, cooks himself a scrambled egg, and settles down on the sofa with *The Essays of E. B. White*. Around noon, he rides his bike to the bookstore. He spends a couple of hours there, just poking around the books. Then he pedals back through the little town, past his house, and to the lake.

When the woman woke up this morning, she got out of bed and went immediately to her easel, where she picked up her pastels and set to work on her painting. After an hour, she is satisfied with the light effect and quits to have breakfast. She dresses quickly and walks to a nearby store to buy shutters for her bathroom. At the store, she meets friends and has lunch with them. Afterward, she wants to be alone and drives to the lake.

Now, the man and the woman stand on the wooden dock, gazing at the lake and the waves on the water. They haven't noticed each other.

The man turns. And so begins the sequence of events informing him of her. Light reflected from her body instantly enters the pupils of his eyes, at the rate of ten trillion particles of light per second. Once through the pupil of each eye, the light travels through an oval-shaped lens, then through a transparent, jellylike substance filling up the eyeball, and lands on the retina. Here it is gathered by one hundred million rod and cone cells.

Cells in the path of reflected highlights receive a great deal of light; cells falling in the shadows of the reflected scene receive very little. The woman's lips, for example, are just now glistening in the sunlight, reflecting light of high intensity onto a tiny patch of cells slightly northeast of back center of the man's retina. The edges around her mouth, on the other hand, are rather dark, so that cells neighboring the northeast patch receive much less light.

Each particle of light ends its journey in the eye upon meeting a retinene molecule, consisting of 20 carbon atoms, 28 hydrogen atoms, and 1 oxygen atom. In its dormant condition, each retinene molecule is attached to a protein molecule and has a twist between the eleventh and fifteenth carbon atoms. But when light strikes it, as is now happening in about 30,000 trillion retinene molecules every second, the molecule straightens out and separates from its protein. After several intermediate steps, it wraps into a twist again, awaiting arrival of a new particle of light. Far less than a thousandth of a second has elapsed since the man saw the woman.

Triggered by the dance of the retinene molecules, the nerve cells, or neurons, respond. First in the eye and then in the brain. One neuron, for instance, has just gone into action. Protein molecules on its surface suddenly change their shape, blocking the flow of positively charged sodium atoms from the surrounding body fluid. This change in flow of electrically charged atoms produces a change in voltage that shudders through the cell. After a distance of a fraction of an inch, the electrical signal reaches the end of the neuron, altering the release of specific molecules, which migrate a distance of a hundred-thousandth of an inch until they reach the next neuron, passing along the news.

The woman, in fact, holds her hands by her sides and tilts her head at an angle of five and a half degrees. Her hair falls just to her shoulders. This information and much, much more is exactingly encoded by the electrical pulses in the various neurons of the man's eyes.

In another few thousandths of a second, the electrical signals reach the ganglion neurons, which bunch together in the optic nerve at the back of the eye and carry their data to the brain. Here, the impulses race to the primary visual cortex, a highly folded layer of tissue about a tenth of an inch thick and two square inches in area, containing one hundred million

neurons in half a dozen layers. The fourth layer receives the
input first, does a preliminary analysis, and transfers the information to neurons in other layers. At every stage, each neuron
may receive signals from a thousand other neurons, combine
the signals—some of which cancel each other out—and dis-
patch the computed result to a thousand-odd other neurons.

After about thirty seconds—after several hundred trillion
particles of reflected light have entered the man's eyes and
been processed—the woman says hello. Immediately, molecules of air are pushed together, then apart, then together,
beginning in her vocal cords and traveling in a springlike
motion to the man's ears. The sound makes the trip from her
to him (twenty feet) in a fiftieth of a second.

Within each of his ears, the vibrating air quickly covers the
distance to the eardrum. The eardrum, an oval membrane
about .3 inch in diameter and tilted fifty-five degrees from the
floor of the auditory canal, itself begins trembling and trans-
mits its motion to three tiny bones. From there, the vibrations
shake the fluid in the cochlea, which spirals snail-like two and
a half turns around.

Inside the cochlea the tones are deciphered. Here, a very
thin membrane undulates in step with the sloshing fluid, and
through this basilar membrane run tiny filaments of varying
thicknesses, like strings on a harp. The woman's voice, from
afar, is playing this harp. Her hello begins in the low registers
and rises in pitch toward the end. In precise response, the thick
filaments in the basilar membrane vibrate first, followed by
the thinner ones. Finally, tens of thousands of rod-shaped bod-

ies perched on the basilar membrane convey their particular quiverings to the auditory nerve.

News of the woman's hello, in electrical form, races along the neurons of the auditory nerve and enters the man's brain, through the thalamus, to a specialized region of the cerebral cortex for further processing. Eventually, a large fraction of the trillion neurons in the man's brain become involved with computing the visual and auditory data just acquired. Sodium and potassium gates open and close. Electrical currents speed along neuron fibers. Molecules flow from one nerve ending to the next.

All of this is known. What is not known is why, after about a minute, the man walks over to the woman and smiles.

Is the Earth Round or Flat?

I PROPOSE THAT there are few of us who have personally verified that the Earth is round. The suggestive globe standing in the den or the Apollo photographs don't count. These are secondhand pieces of evidence that might be thrown out entirely in court. When you think about it, most of us simply believe what we hear. Round or flat, whatever. It's not a life-or-death matter, unless you happen to live near the edge.

A few years ago I suddenly realized, to my dismay, that I didn't know with certainty whether the Earth was round or flat. I have scientific colleagues, geodesists they are called, whose sole business is determining the detailed shape of the Earth by fitting mathematical formulae to *someone else's* measurements of the precise locations of test stations on the Earth's surface. And I don't think those people really know either.

Aristotle is the first person in recorded history to have given proof that the Earth is round. He used several different arguments, most likely because he wanted to convince others as well as himself. A lot of people believed everything Aristotle said for nineteen centuries.

His first proof was that the shadow of the Earth during a lunar eclipse is always curved, a segment of a circle. If the Earth were any shape but spherical, the shadow it casts, in some orientations, would not be circular. (That the normal phases of the moon are crescent-shaped reveals the moon is round.) I find this argument wonderfully appealing. It is simple and direct. What's more, an inquisitive and untrusting person can knock off the experiment alone, without special equipment. From any given spot on the Earth, a lunar eclipse can be seen about once a year. You simply have to look up on the right night and carefully observe what's happening. I've never done it.

Aristotle's second proof was that stars rise and set sooner for people in the east than in the west. If the Earth were flat from east to west, stars would rise as soon for occidentals as for orientals. With a little scribbling on a piece of paper, you can see that these observations imply a round Earth, regardless of whether it is the Earth that spins around or the stars that revolve around the Earth. Finally, northbound travelers observe previously invisible stars appearing above the northern horizon, showing the Earth is curved from north to south. Of course, you do have to accept the reports of a number of friends in different places or be willing to do some traveling.

Aristotle's last argument was purely theoretical and even philosophical. If the Earth had been formed from smaller pieces sometime in the past (or *could* have been so formed), its pieces would fall toward a common center, thus making a sphere. Furthermore, a sphere is clearly the most perfect solid shape. Interestingly, Aristotle placed as much emphasis on this last argument as on the first two. Those days, before the modern "scientific method," observational check wasn't required for investigating reality.

Assuming for the moment that the Earth is round, the first person who measured its circumference accurately was another Greek, Eratosthenes (276–195 B.C.). Eratosthenes noted that on the first day of summer, sunlight struck the bottom of a vertical well in Syene (modern Aswan), Egypt, indicating the sun was directly overhead. At the same time in Alexandria, 5,000 stadia distant, the sun made an angle with the vertical equal to 1/50 of a circle. (A stadium equaled about a tenth of a mile.) Since the sun is so far away, its rays arrive almost in parallel. If you draw a circle with two radii extending from the center outward through the perimeter (where they become local verticals), you'll see that a sun ray coming in parallel to one of the radii (at Syene) makes an angle with the other (at Alexandria) equal to the angle between the two radii. Therefore Eratosthenes concluded that the full circumference of the Earth is 50 × 5,000 stadia, or about 25,000 miles. This calculation is within 1 percent of the best modern value.

For at least 600 years educated people have believed the Earth is round. At nearly any medieval university, the quadrivium was standard fare, consisting of arithmetic, geometry, music, and astronomy. The astronomy portion was based on the *Tractatus de Sphaera Mundi*, a popular textbook first published at Ferrara, Italy, in 1472, and written by a thirteenth-century, Oxford-educated astronomer and mathematician, Johannes de Sacrobosco. The *Sphaera* proves its astronomical assertions, in part, by a set of diagrams with movable parts, a graphical demonstration of Aristotle's second method of proof. The round Earth, being the obvious center of the universe, provides a fixed pivot for the assembly. The cutout figures of the sun, the moon, and the stars revolve about the Earth.

By the year 1500, twenty-four editions of the *Sphaera* had appeared. There is no question that many people *believed* the Earth was round. I wonder how many *knew* this. You would think that Columbus and Magellan might have wanted to ascertain the facts for themselves before waving good-bye.

To protect my honor as a scientist, someone who is supposed to take nothing for granted, I set out with my wife on a sailing voyage in the Greek islands. I reasoned that at sea I would be able to calmly observe landmasses disappear over the curve of the Earth and thus convince myself, firsthand, that the Earth is round.

Greece seemed a particularly satisfying place to conduct my experiment. I could sense those great ancient thinkers looking on approvingly, and the layout of the place is perfect. Hydra rises about 2,000 feet above sea level. If the Earth has a radius of 4,000 miles, as they say, then Hydra should appear to sink down to the horizon from a distance of about fifty miles, somewhat less than the distance we were to sail from Hydra to Kea. The theory was sound and comfortable. At the very least, I thought, we would have a pleasant vacation.

As it turned out, that was all we got. Every single day was hazy. Islands faded from view at a distance of only eight miles, when the land was still a couple of degrees above the horizon. I learned how much water vapor was in the air but nothing about the curvature of the Earth.

I suspect that there are quite a few items we take on faith, even important things, even things we could verify without much trouble. Is the gas we exhale the same as the gas we inhale? (Do we indeed burn oxygen in our metabolism, as they say?) What is our blood made of? (Does it indeed have red and white "cells"?) These questions could be answered with a balloon, a candle, and a microscope.

When we finally do the experiment, we relish the knowledge. At one time or another, we have all learned something for ourselves, from the ground floor up, taking no one's word for it. There is a special satisfaction and joy in being able to tell somebody something you have pieced together from scratch, something you really know. I think that exhilaration is a big reason why people do science.

Someday soon, I'm going to catch the Earth's shadow in a lunar eclipse, or go to sea in clear air, and find out for sure if the Earth is round or flat. Actually, the Earth is reported to flatten at the poles, because it rotates. But that's another story.

If Birds Can Fly,
Why, Oh Why, Can't I?

HUMAN PHYSICAL CAPACITY is greatly restricted by natural laws, nowhere better illustrated than by our inability, despite vigorous and patient flapping of the arms, to fly. But the problem here is not simply the lack of wings. Scale up a pheasant to the size of a man and it would plummet to earth like a rock. Or consider Icarus. In the very plausible picture of him in my childhood mythology book, each attached wing equals his height and is about one quarter as wide—not unlike the graceful proportions of a swallow. Unfortunately, to fly with those wings the boy would have to beat his arms at one and a half horsepower, four times the maximum sustained output of an athletic human being. Icarus and Daedalus may have been willing to utterly exhaust themselves in their aerial escape from Crete, but most of us would like to go with better equipment.

Weight, shape, and available power all play a part in the science of flying. Let us begin with the most obvious requirement to fly: a lifting force must counterbalance the weight of the animal in question. That lift is provided by air. Air has weight and, at sea level, pushes equally in all directions with a pressure just under fifteen pounds per square inch of surface. To achieve lift, an animal must manage to reduce the air pressure on its top surface, thereby creating a net pressure pushing upward from below. Birds and airplanes do this with properly formed wings and forward motion. The curvature and trailing edge of a wing force the air to flow more rapidly over its top side than its bottom. This causes a net upward pressure in proportion to the air density and to the square of the forward speed, a basic law of physics deriving from the conservation of energy. Thus with every doubling of the speed comes a quadrupling of the lift pressure. No motion, no net lift pressure. Likewise, birds couldn't fly on the moon, where the air density is essentially zero. (Under the moon's reduced gravity, however, creatures could jump six times as high as on Earth, which might be a happy substitute.)

Once you've got your lift pressure of so many pounds per square inch, you want to have out as many square inches of wing as practical. For example, a lift pressure of a hundredth pound per square inch (obtained by flying at about thirty-five miles per hour) pushing on a wing area of 400 square inches will yield a total lift force of four pounds, enough to buoy the weight of the average bird. There is a convenient tradeoff here: the necessary lift force can be had with less wing area if the animal increases its forward speed, and vice versa. Birds capitalize on this option according to their individual needs. The great blue heron, for example, has long, slender legs for wading and must fly slowly in order not to break them on landing. Consequently, herons have a relatively large wingspan. Pheasants, on the other hand, maneuver in underbrush and would find large wings cumbersome. To remain airborne with their relatively short and stubby wings, pheasants fly fast. Illustrating with some actual numbers, which I got by telephone from a helpful man at the Audubon Society who happened to have the birds in his office, an average great blue heron weighs in at six and a half pounds and projects a wing area of about 800 square inches, while a typical pheasant has three times the weight to wing area. However, the pheasant flies at a brisk fifty miles per hour, twice the speed of the heron.

How a bird propels itself forward, without propellers, is not obvious. This mystery was clarified in the early nineteenth century by Sir George Cayley, father of the modern airplane. (Leonardo Da Vinci spent years studying the art of flying and may well have understood the propulsion of birds, but his notes went undiscovered until a hundred years ago and, as typical, were left unfinished—although he did, as legend has it, launch one of his pupils from Mount Cecere in a flying contraption, which promptly crashed.) Birds, in fact, do have propellers, in the form of specially designed feathers in the outer halves of their wings. These feathers, called primaries, change their shape and position during a wing beat. On the downstroke they move downward and forward; on the upstroke, upward and backward. The primary feathers, operating on the same physical principles as the rest of the wing, produce their lift in the forward rather than upward direction.

Flying, like other physical activities, costs energy. A frictionless bird, having attained level flight and satisfied with its course, could glide forever, without moving a muscle. All the flapping, and expense of energy, is made necessary by air drag. Depending on a craft's aerodynamic design, the drag force is something like a twentieth of the lift force. To counteract drag, a cruising heron must pay out energy at the grudging rate of a fiftieth of a horsepower, leaving behind its calories in the form of stirred up pockets of air. Heavier birds of the same proportions have to use even more power for each pound of weight. Quadruple a bird's dimensions in every direction, keeping the shape identical, and its weight and volume increase by 64 times, while the power required to fly is 128 times larger. The only way around this law is to change shape. For example, if you hold the total volume (and weight) fixed, but increase the wing area four times, you can fly with half the power. For the long flights in migration, birds save power by flying together in formation, each member in the rear partially boosted by the rising air current trailing the next bird up and taking turns in the lead positions. But for solo flight, weight and shape inescapably determine the power needed to fly. These are the facts of life for aviation.

We now turn to biology, where we find that, pound for pound, living creatures are highly inefficient at producing useful power compared to internal combustion engines. A human being can maintain a maximum mechanical power output of only about one two-hundredth that of an engine of the same weight. For biology to reach twelve horsepower, the output of the little engine on the Wright brothers' 1903 airplane, requires the services of an elephant.

The predicament of limited biological power lessens, however, as we proceed to smaller and smaller size. Lighter animals have more power for each pound of weight than heavier ones. Begin with a 450-pound horse, which has at its disposal one horsepower. Now reduce the weight of the animal. For every fifty percent reduction of weight, it is found that the power available for work diminishes by only forty percent. By the time you're down to less than an ounce, you realize the 4,000 mice that weigh the same as a man have nine times the total power. Embarrassing perhaps, but not unexpected. Unlike most engines and machines, the muscles in animals generate more heat than useful work. Since heat production cannot be sustained at a greater rate than the animal can cool, and cooling is generally accomplished via the skin surface, animals will produce heat, and mechanical power, approximately in proportion to their surface area. The ratio of power output to weight, therefore, is close to the ratio of surface area to volume. It is then simple mathematics that smaller objects have greater surface area to volume than larger ones. (Very small animals do have their own drawbacks, like the obligation to eat for most of the day, but that's another story.)

Now, as the power needed to fly increases more rapidly with increasing weight than the power generally available—unless some spectacular change in body shape is effected—

lightweight creatures clearly have the edge for flying. Nature seems quite appreciative of this struggle between physics and biology. Although birds have been experimenting with flight for one hundred million years, the heaviest true flying bird, the great bustard, rarely exceeds thirty-two pounds. The larger, gliding birds such as vultures, lifted by rising hot air columns, do not carry their full weight. The 300-pound ostrich never leaves the ground, apparently having chosen sheer bulk rather than flight for defense.

Never having seen a 200-pound bird aloft, the British industrialist Henry Kremer must have felt his money safe for a long time when in 1959 he offered a prize of £5,000 for human-powered flight. In 1973, after many serious but unsuccessful attempts by Britain, Japan, Austria, and Germany, the Kremer Prize was increased to the equivalent of $86,000. According to the strict rules of the offer, as set up by the Royal Aeronautical Society of England, a winning flight would have to traverse a figure eight around two pylons half a mile apart, never touching earth, and cross the start and finish point at least 10 feet above the ground. And, of course, the human pilot would have to furnish the power.

On August 23, 1977, in Shafter, California, an athletic young man climbed into a fragile, ungainly craft named the Gossamer Condor, strapped his feet to a pair of bicyclelike pedals connected to a propeller, and captured the prize. The flight lasted about seven and a half minutes.

What Paul MacCready, the designer of the craft, had done

was to create an extraordinarily lightweight structure of enormous wingspan. To keep the wing as light as possible, it was fashioned from Mylar stretched over aluminum struts, with piano wire for bracing and cardboard for the wing's leading edge. The entire craft, including fuselage and wing, weighs seventy pounds. To this you have to add the weight of Bryan Allen, 135 pounds. Allen is approximately six feet tall; his wing was ninety-six feet long and ten feet wide. Never has Mother Nature conceived a flying creature remotely approaching such disproportions. For numerical comparison, the pheasant has a wing area per mean body area of just over one, the heron has five, and the great bustard has thirteen. The Gossamer Condor (with pilot) has a wing area per mean body area of ninety.

In many ways, human beings circumvented the difficulties of aviation long ago, at Kitty Hawk. And internal combustion engines date back even earlier. But in our dreams, when we soar into the air to escape danger or to simply bask in our strength, we fly as birds, self-propelled. It may be awkward to imagine ourselves installed with one hundred feet of wing, but that's what Nature asks, to fly like a bird.

Students and Teachers

In the fall of 1934, one year after his Ph.D., John Archibald Wheeler traveled to Copenhagen to study with the great atomic physicist Niels Bohr. At his Institute for Theoretical Physics, a house-sized building on Blegdamsvej 15, Bohr had created a scientific "school" in which the daily stimulation from brilliant seminars and disturbing new ideas could dismast slow thinkers. Among the students who had held up well were Felix Bloch, Max Delbrück, Linus Pauling, and Harold Urey, all future Nobel Prize winners like their teacher. As Wheeler arrived at the institute on bicycle one morning, he noticed a workman tearing down the vines that had grown thickly over the gray stucco front. On closer view, he saw it was Bohr himself, "following his usual modest but direct approach to a problem." Thus began Wheeler's tutelage.

I am, through Wheeler, a great-grandstudent of Bohr. I had forgotten this fact until a recent venture to the Boston studio of painter Paul Ingbretson, who immediately announced his own pedagogical descent from R. H. Ives Gammell, a student of William Paxton, a student of the academic painter Jean-Léon Gérôme. In the art world it is commonly said that the days of the master and apprentice tradition ended two centuries ago, that the classical method of severe and thorough training has been lost, except by a handful of painters. Ingbretson, thirty-four years old, is one of that handful. He treasures the technique and style and wisdom garnered from his teachers and tries to give away some of it to his pupils here in the Fenway Studios, where Paxton worked from 1905 to 1914. What he learned from Gammell and Paxton cannot be written down. He is a living painting, full of their brush strokes and visions. In science, such personal inheritance receives less currency, following the idea that a cut-and-dried objectivity outlives questions of style. You rarely witness a scientist exhibiting his pedagogical lineage. Yet without a good teacher, a young student of science could read a row of textbooks stretching to the moon and not learn how to practice the trade. Exactly what is it, in this age of massive information storage and retrieval, that you can't learn from a book?

"Squint your eyes; squint your eyes," Ingbretson was admonishing one of his students. By squinting the eyes as you study your subject, minor details fade, leaving only the highlights, the dominant lights and darks. Ingbretson's charges, with their easels and paper and charcoal, were clustered around a classical marble bust, lit from windows rising to the sixteen-foot ceiling of the studio. "It's all a question of learning to see," Ingbretson was saying. This phrase "learning to see" was one Gammell often used. It typifies the method of painting from nature practiced by the early-twentieth-century Boston school, the blend of the exacting academic style with impressionism.

Wheeler, now eighty-five, had his own method of learning how to see, which he taught to my teacher Kip Thorne: "If you're having trouble thinking clearly, imagine programming a computer to solve your problem. After mentally automating the necessary logic, step by step, you can then dispense with the computer." On occasion, marching through a problem in such a fashion will lead to an unexpected contradiction, and this is where the fun really begins. Wheeler loves to teach physics in terms of paradoxes, a habit he picked up from Bohr. In the 1920s and 1930s—when quantum mechanics was in its infancy and physicists were slowly adjusting to the strange fact that an electron behaves both as a localizable particle and as a wave, spread over many places at once—Bohr realized that several apparently conflicting views can be equally essential for understanding some phenomena. A student doesn't get this kind of thinking from books. Wheeler recalls Bohr's usual method of explanation as a one-man tennis match, in which each hit of the ball would be some telling contradiction to previous results, raised by a new experiment or theory. After each hit Bohr would run around to the other side of the court quickly enough to return his own shot. "No progress without a paradox." The worst thing that could happen in a visitor's seminar was the absence of surprises, after which Bohr had to utter those dreaded words, "That was interesting."

I was slowly circling an odd still-life arrangement in the clutter of Ingbretson's studio—an upright porcelain plate with a diagonal pattern, a bowl, a matchbox, a bit of dried flowers. On an easel nearby was an extremely effective rendering by one of Ingbretson's advanced pupils, clearly taken from the pale still life in front of me, yet more interesting somehow. Then I slipped around to the precise viewing angle of the drawing and suddenly the objects on the table leaped out at me in a wonderful way. "Some artists," said Ingbretson, "will arrange and rearrange a still life for hours until they find just the right grouping and viewing angle." You look at it from the wrong direction and all you've got is a collection of junk. "Sometimes reality isn't enough. I was once doing a still life in Gammell's class, and we had meticulously chosen and arranged the objects beforehand. After I was finished Gammell stared at my work for a few minutes and then told me to draw in a nonexistent knickknack in the corner. It turned out he was right."

Graduate students in science, unanchored to a knowledge-able thesis adviser, have wasted years circling around for a good project. Every so often an application comes in from a Third World student wanting a research position abroad, and you can tell he's highly competent mathematically and he's been combing the journals equation by equation, but his teachers are isolated from mainstream research, and he has no clue as to what projects are worth working on. The Nobel

Prize-winning Soviet physicist Lev Landau kept a notebook of about thirty important unsolved problems, which he would show to students if and when they successfully passed a barrage of tests known affectionately as the Landau Minimum. Significant research projects in science are often no more difficult than insignificant ones. Projects out of Landau's notebook had guaranteed significance.

As a student, you could always tell which projects Thorne was hot on, because the hallway near his office was lined with framed wagers between himself and other scientific eminences. "Kip Steven Thorne wagers S. Chandrasekhar that rotating black holes will prove to be stable. K. S. T. places forward a year's subscription to *The Listener*. S. C. places forward a year's subscription to *Playboy*." And so on. Thorne, red-bearded and wiry, would sit quietly in his office filling pages with equations, while passing students contemplated those wagers in the hall and were set on fire.

Beethoven and Czerny and Liszt, Socrates and Plato and Aristotle, Verrocchio and Leonardo, Pushkin and Baryshnikov. As we stood in the studio, Ingbretson walked over to a pupil who had succeeded in putting down three questionable lines in the last hour and told her to start from scratch. Ingbretson's own teacher demanded a lot from his students and didn't mind a little humiliation to get a point across. One day while a younger Ingbretson was smugly reflecting on his painting, Gammell, a bald, wizened man with the head of a bulldog, standing not much over five feet, took Ingbretson by his pinky, led him around the room to some white paint, dipped the little finger in the paint, then led him back to the canvas and applied the finger to a strategic spot. "There," said Gammell, "now you've got the highlight." Hans Krebs, winner of the 1953 Nobel Prize in medicine or physiology, a student of Nobelist Otto Warburg, a student of Nobelist Emil Fischer, wrote that scientists of distinction, above all, "teach a high standard of research. We measure everything, including ourselves, by comparisons; and in the absence of someone with outstanding ability there is a risk that we easily come to believe we are excellent. . . . Mediocre people may appear big to themselves (and to others) if they are surrounded by small circumstances. By the same token big people feel dwarfed in the company of giants, and this is a most useful feeling. . . . If I ask myself how it came about that one day I found myself in Stockholm, I have not the slightest doubt that I owe this good fortune to the circumstance that I had an outstanding teacher at the critical stage of my scientific career."

Rulers and plumb bobs appear regularly in Ingbretson's studio. A plumb bob is a weight attached to a string and, when freely hung in the earth's gravitational pull, gives an unerring reading of the vertical direction. Rulers and plumb bobs serve as invaluable tools for getting proportion and angles exactly right. This old tradition of expert draftsmanship was gleaned from Gammell, who learned it from Paxton. Paxton's portraits are stunning in their precision, with a reality and sensuality far exceeding any photograph. Of Paxton, Gammell once wrote: "His unsurpassed visual acuity combined with great technical command enabled him to report his impressions with astounding veracity."

One of Ingbretson's students was struggling with angles in her drawing of the marble bust. Lines were going awry and wandering aimlessly. The sacred plumb bob wasn't working. "Aha," he offered, "your paper's gotten tilted."

Learning good draftsmanship requires constant feedback between teacher and student, Ingbretson explained. But good draftsmanship isn't enough. After mastering technique, you then must decide what to emphasize on the canvas. This tricky combination of formal method and individual impression has the flavor of the balance between mathematical rigor and physical intuition required in science. Thorne believes a feeling for such balance is one of the crucial things he learned from Wheeler. "Many scientists move at a snail's pace because they are too mathematical and don't know how to think physically. And vice versa for people too sloppy in their mathematics." Consider, for example, a quantitative description of marbles rolling around on a floor with holes in it. You try to derive an equation that tells how the number of marbles decreases in time. A most useful check of that equation is to set the hole size to a small number, which should yield the result that you don't lose any of your marbles. Or else the equation is wrong. This check wouldn't naturally occur to you unless you have a physical picture in your head of marbles rolling around and falling, one by one, through the holes. The mathematical equation itself, right or wrong, is quite content to stare back with an unrevealing jumble of its marble-conserving and marble-nonconserving parts.

Niels Bohr was a barrel-chested man, a football hero in his

younger days. He was also gentle, and made his penetrating points in a soft voice. Bohr had many ideas he never tried to copyright. Likewise, his student John Wheeler, who quietly introduced numerous seminal ideas in physics, who performed an important but little-known role advising the DuPont company during the Manhattan Project. Personal style can be inherited. Wheeler's student, Kip Thorne, has always bent over backward to give credit to other scientists. He begins seminars by attributing most of his results to particular students. Modesty, and its opposite, set the tone of a research group.

Ghirlandajo and Michelangelo, Koussevitsky and Bernstein, Lastman and Rembrandt, Fermi and Bethe, Luria and Watson. Of the 286 Nobel laureates named between 1901 and 1972, forty-one percent had a master or senior collaborator who was also a Nobelist. Many Nobelists have surrounded themselves with spirited schools of students. A cluster of apprentices seems to generate, en masse, the necessary speed for takeoff. Among the great recent masters in physics were Thomson and Rutherford in England, Landau and Zel'dovich in the former Soviet Union, Bohr in Denmark, Fermi and Oppenheimer and Alvarez in the United States—all with large research groups that spawned other eminent scientists. At Caltech, Thorne has always insisted on cloistering his half-dozen students within adjacent rooms, with an unwritten rule that office and lab doors remain open. Someone, in a group of creative people working together, is usually quivering at the edge of discovery, and the vibrations spread.

Gazing out from a photograph of the Boston Museum's 1913 life-drawing class is a mustached, steady-eyed Paxton, sitting among his seventeen students. On the front left is Gammell, twenty years old, wearing an overcoat and a full head of hair. His expression is serious. The other pupils stand or sit, some wear elegant suits and others short sleeves and smocks, some look frightened and others bored, but they lean into each other with hands on shoulders, and there is electricity in the air.

The light was fading from the tall windows in Ingbretson's studio and his pupils were packing up their materials. "You know, Gammell wasn't perfect. His gestures were forced. Look at that arm." Ingbretson held up an illustration in a book of Gammell's paintings. "That's unnatural. It took me a while to see it. I was relieved."

Nothing is more bracing for students than to discover the fallibility of their exalted teachers. Students, God knows, are brimming with their own human weaknesses, and if their great mentors can make mistakes, well then, anything might happen. Thorne remembers that during his second year of graduate school, Wheeler stuck by some erroneous statements about black holes. The realization of Wheeler's errors provided its own kind of inspiration. When Wheeler was in Copenhagen in 1934, he sought Bohr's assessment of some calculations on the so-called dispersion theory, extending it from applications where particles move slowly to applications where they move at nearly the speed of light. Bohr was skeptical of Wheeler's work and discouraged publication. Bohr was wrong. Perhaps, in the end, our own imperfection is the most vital thing we learn from teachers. At the dedication of the giant statue of Einstein in Washington fifteen years ago, Wheeler said, "How can we best symbolize that science reaches after the eternal? . . . Not by a pompous figure on a pedestal. No, a figure over which children can crawl. . . . "

Time Travel and Papa Joe's Pipe

When astronomers point their telescopes to the nearest large galaxy, Andromeda, they see it as it was two million years ago. That's about the time Australopithecus was basking in the African sun. This little bit of time travel is possible because light takes two million years to make the trip from there to here. Too bad we couldn't turn things around and observe Earth from some cozy planet in Andromeda.

But looking at light from distant objects isn't real time travel, the in-the-flesh participation in past and future of Mark Twain's Connecticut Yankee or H. G. Wells's Time Traveler. Ever since I've been old enough to read science fiction, I've dreamed of time traveling. The possibilities are staggering. You could take medicine back to fourteenth-century Europe and stop the spread of plague, or you could travel to the twenty-third century, where people take their annual holidays in space stations.

Being a scientist myself, I know that time travel is quite unlikely according to the laws of physics. For one thing, there would be causality violation. If you could travel backward in time, you could alter a chain of events with the knowledge of how they would have turned out. Cause would no longer always precede effect. For example, you could prevent your parents from ever meeting. Contemplating the consequences of that will give you a headache, and science-fiction writers for decades have delighted in the paradoxes that can arise from traveling through time.

Physicists are, of course, horrified at the thought of causality violation. Differential equations for the way things should behave under a given set of forces and initial conditions would no longer be valid, since what happens in one instant would not necessarily determine what happens in the next. Physicists do rely on a deterministic universe in which to operate, and time travel would almost certainly put them and most other scientists permanently out of work.

But even within the paradigms of physics, there are some technical difficulties for time travel, over and above the annoying fact that its existence would altogether do away with science. The manner in which time flows, as we now understand it, was brilliantly elucidated by Albert Einstein in 1905. First of all, Einstein unceremoniously struck down the Aristotelian and Newtonian ideas of the absoluteness of time, showing that the measured rate at which time flows can vary between observers in relative motion with respect to each other. So far, this looks hopeful for time travel.

Einstein also showed, however, that the measured time order of two events could not be reversed without relative motions exceeding the speed of light. In modern physics the speed of light, 186,000 miles per second, is a rather special speed; it is the propagation speed of all electromagnetic radiation in a vacuum, and appears to be nature's fundamental speed limit. From countless experiments, we have failed to find evidence of anything traveling faster than light.

There is another possible way out. In 1915 Einstein enlarged his 1905 theory, the Special Theory of Relativity, to include the effects of gravity; the later theory is imaginatively named the General Theory of Relativity. Both theories have remarkably survived all the experimental tests within our capability. According to the General Theory, gravity stretches and twists the geometry of space and time, distorting the temporal and spatial separation of events.

The speed of light still cannot be exceeded locally—that is, for brief trips. But a long trip might sneak through a short cut in space created by gravitational warping, with the net result that a traveler could go between two points by one route in less time than light would require by another route. It's a little like driving from Las Vegas to San Francisco, with the option of a detour around Death Valley. In some cases, these circuitous routes might lead to time travel, which would indeed raise the whole question of causality violation.

The catch is that it is impossible to find any concrete solutions of Einstein's equations that permit time travel and are at the same time well behaved in other respects. All such proposals either require some unattainable configuration of matter, or else have at least one nasty point in space called a "naked singularity" that lies outside the domain of validity of the theory. It is almost as if General Relativity, when pushed toward those circumstances in which all of physics is about to be done away with, digs in its heels and cries out for help.

Still, I dream of time travel. There is something very personal about time. When the first mechanical clocks were invented, marking off time in crisp, regular intervals, it must have surprised people to discover that time flowed outside their own mental and physiological processes. Body time flows at its own variable rate, oblivious to the most precise hydrogen maser clocks in the laboratory.

In fact, the human body contains its own exquisite timepieces, all with their separate rhythms. There are the alpha waves in the brain; another clock is the heart. And all the while tick the mysterious, ruthless clocks that regulate aging.

Nowhere is the external flow of time more evident than in the space-time diagrams developed by Hermann Minkowski, soon after Einstein's early work. A Minkowski diagram is a graph in which time runs along the vertical axis and space along the horizontal axis. Each point in the graph has a time coordinate and a space coordinate, like longitude and latitude, except far more interesting. Instead of depicting only where something is, the diagram tells us when as well.

In a Minkowski diagram, the entire life history, past and future, of a molecule or a man is simply summarized as an unbudging line segment. All this on a single piece of paper. There is something disturbingly similar about a Minkowski diagram and a family tree, in which several generations, from long dead relatives to you and your children, move inevitably downward on the page. I have an urgent desire to tamper with the flow.

Recently, I found my great-grandfather's favorite pipe. Papa Joe, as he was called, died more than seventy years ago, long before I was born. There are few surviving photographs or other memorabilia of Papa Joe. But I do have this pipe. It is a fine old English briar, with a solid bowl and a beautiful straight grain. And it has a silver band at the base of the stem, engraved with three strange symbols. I should add that in well-chosen briar pipes the wood and tobacco form a kind of symbiotic relationship, exchanging juices and aromas with each other, and the bowl retains a slight flavor of each different tobacco smoked in the pipe.

Papa Joe's pipe had been tucked away in a drawer somewhere for years, and was in good condition when I found it. I ran a pipe cleaner through it, filled it with some tobacco I had on hand, and settled down to read and smoke. After a couple of minutes, the most wonderful and foreign blend of smells began wafting from the pipe. All the various tobaccos that Papa Joe had tried at one time or another in his life, all the different occasions when he had lit his pipe, all the different places he had been that I will never know—all had been locked up in that pipe and now poured out into the room. I was vaguely aware that something had got delightfully twisted in time for a moment, skipped upward on the page. There *is* a kind of time travel to be had, if you don't insist on how it happens.

In His Image

I RECENTLY CAME ACROSS a collection of scientific studies on the search for extraterrestrial intelligence, published a few years ago by the National Aeronautics and Space Administration. The book's foreword was written by Theodore M. Hesburgh, who is president of the University of Notre Dame and a Catholic theologian. Hesburgh recalls being asked (by a surprised lawyer) how a religious person such as himself could legitimately accept the possibility of other inhabited worlds out there in space. He answered: "It is precisely because I believe theologically that there is a being called God, and that He is infinite in intelligence, freedom, and power, that I cannot take it upon myself to limit what He might have done." Writing on the same proposition seven hundred years earlier, and speaking for many of the intellectuals of his day, the great theologian and philosopher St. Thomas Aquinas took exactly the opposite position: "This world is called one by the unity of order . . . all things should belong to one world." For Aquinas, who spent his life trying to reconcile faith with reason, God's omnipotence and goodness were better illustrated by a single, perfect world than by many, necessarily imperfect worlds.

The Reverend Hesburgh—and the many of us who share his ease with the possibility of other worlds—has not come to his point of view by any certain scientific evidence. Despite a great deal of searching, no life of any kind has turned up on Mars; no extraterrestrial communications have yet been detected from outer space; few planets have yet been found outside our solar system. What happened between Aquinas's day and ours was a revolution in how we think of ourselves in the grand scheme of things. It was a revolution that occurred mainly in the seventeenth century. It was part of the birth of modern science, but it went far beyond science; it was part of the beginnings of Protestantism and the victory of natural theology over Scripture, but it went far beyond religion; it was part of the French Enlightenment and the Age of Reason. The extraterrestrial question, the question of whether our minds are necessarily unique in the universe, penetrates to the deepest roots of our culture and our identity as human beings.

The notion of other worlds goes back at least as far as the Greek atomists and their view that space is filled by an infinite number of similar atoms, all obeying the same natural laws. In such a philosophy, whatever happened on earth would have been repeated all through the cosmos. Aristotle, however, strongly disagreed with this picture. All things, according to Aristotle, were composed of five elements: earth, air, fire, water, and ether—and each element had its "natural place." The natural place of "earth" was at the center of the universe, and all earthlike particles anywhere in the universe would fall to that place. The natural place of ether was in the outermost heavens, where it made up the stars. Water, air, and fire had intermediate locations. Aristotle's intellectual grip through the centuries was powerful. For one thing, his view of an earth-centered cosmos appeals strongly to common sense. Standing outside on a starry night, it's easy to believe that the universe revolves around us and that those distant points of light are made of some non-earthly material. Everything in its place. Aquinas made it his business to reconcile Christianity with Aristotelian philosophy whenever possible.

Aristotle aside, there were strong religious and emotional reasons for rejecting the possibility of other worlds. Only one earth is mentioned in Scripture, for example. Perhaps more importantly, the whole tone of the Bible suggests a comfortable, personal relationship between man and God. God watches over us. In medieval times, it was also widely believed that the universe was created principally for us and our use. Life is baffling as it is. Who wants to live out her days with dubious status, in a cosmos with uncertain purpose? People were not in a hurry to give these things up.

Even so, not everyone was able to wed faith and reason so peacefully as Aquinas. God's omnipotence and creative power would be diminished by being limited to a single world, pronounced the bishop of Paris in 1277. This was a forceful argument in favor of other worlds (essentially the same as Hesburgh's in the NASA book), and it appeared frequently. Theologically, it wasn't clear whether a plurality of worlds would enhance or diminish God's glory. For the next few hundred years, the possibility of other worlds was hotly contested among intellectuals.

In 1543, a critical new element was added to the debate, a scientific element. The *Revolutions of the Heavenly Spheres* of Copernicus was published, announcing that the astronomical data were best fit by placing the sun, rather than the earth, at the center of the solar system. Copernicus didn't speculate on whether Earth's sister planets were earthlike, not to mention inhabited, but the writing was on the wall. This was the beginning of many contributions of science to the question of other worlds, as well as the beginning of modern science itself. In the next century and a half, Galileo's telescopic sightings of irregular mountains on the moon, Kepler's observations of the sudden appearance of a "new" star in the sky where none was before, and Newton's law of universal gravity would all become ammunition in support of the possibility of other worlds.

But it is hard to believe that these technical developments by themselves were the force that eventually turned people's opinions. What I find revealing in this regard is that many of the scientific arguments of this period wobbled on shaky ground, many smelled quaintly of old-fashioned human prejudice and egotism, and all of them invoked the deity in some form or another. The obvious proving ground for the existence of extraterrestrial life was the moon, being closest at hand. To make lunar conditions more comfortable, Galileo (erroneously) hypothesized that the smooth periphery of the moon seen in his telescope was caused by a lunar atmosphere and that the dark spots were lunar oceans. The great German astronomer Johannes Kepler (erroneously) concluded that the shapes and arrangements of the lunar hollows were evidence of architecture by intelligent creatures—so similar in flavor to Percival Lowell's famous "observations" of artificial canals on Mars at the turn of our own century. Although a strong advocate of life on the moon, Kepler took pains to point out that, according to his calculations, our sun was the most luminous (and therefore the noblest) star in the Milky Way. This erroneous result may, perhaps, be understood in light of his comments in *Kepler's Conversation with Galileo's Sidereal Messenger,* 1610:

> [I]f there are globes in the heaven similar to our earth, do we vie with them over who occupies the better portion of the universe? For if their globes are nobler, we are not the noblest of rational creatures. Then how can all things be for man's sake? How can we be the masters of God's handiwork?

Kepler was straddling the fence. Other inhabited worlds in our solar system were all right, but our particular setup was still the best and the brightest. And the astronomer Thomas Wright, in his *Original Theory or New Hypothesis of the Universe* (1750), brandishes his authority as a scientist to support the plurality of worlds, but states up front that "the glory of the Divine Being of course must be the principal object in view" and uses that object in his construction of the cosmos. A lot more than science was at the bottom of these convictions.

During this same period, theology itself was changing dramatically, especially in England. It was increasingly believed that God revealed Himself more in His natural works than in Scripture. Nature was celebrated. This shift in view eventually showed up in the philosophy of Rousseau, in the nature poems of Coleridge and Wordsworth, and in the landscape paintings of Turner and Constable. It was also brought to bear on the question of other worlds. In 1638 the Protestant clergyman John Wilkins, who later became an Anglican bishop, bravely argued that the lack of mention of other worlds in the Bible did not forbid their existence. Some fifty years later, the English theologian Richard Bentley carried the new "natural theology" to its ultimate implications for man:

> [W]e need not nor do not confine and determine the purposes of God in creating all mundane bodies, merely to human ends and uses. . . . All bodies were formed for the sake of intelligent minds: and as the Earth was principally designed for the being and service and contemplation of men; why may not all other

planets be created for the like uses, each for their own
inhabitants which have life and understanding?

And in Milton's *Paradise Lost* (1667), the archangel Raphael
answers Adam's cosmological questions in this way:

> . . . other suns perhaps
> With their attendant Moons thou wilt descry . . .
> Stor'd in each Orb perhaps with some that live.
> For such vast room in Nature unpossesst
> By living Soul, . . . is obvious to dispute . . .

Theological considerations were appearing here in a different
garb than before. God was still a powerful and good force, but
something had changed underneath all the dressing.

As stated so well by Bentley, humankind was becoming
humble—at least intellectually. A landmark in the new humil-
ity was Descartes' *Principles of Philosophy* (1644), the most
comprehensive study of knowledge since Aristotle's, and an
enormous influence on modern thought. In *Principles*,
Descartes thinks about everything from the nature of thought
itself, to the five human senses, to the movement of projectiles,
the behavior of fluids, sunspots, the mechanisms of tides, the
nature of mind, and the soul. Although this may not seem like
the undertaking of a modest man, Descartes places great stock
in that quality as a basis for reasoning. In the third part of his
work, before launching into cosmology, Descartes warns us
not to presume too much in understanding God's purposes,

and then suggests that those purposes are most likely not all for our benefit. Furthermore, and throughout Descartes' vast, nonanthropocentric cosmos, nature follows universal laws. Nature is a single, mechanical system, in which fluids and sunspots dance to the same rules here as they do everywhere in the cosmos. Descartes' ideas drifted like pipe smoke through the salons of Holland, France, Germany, and England. Although the details of his science were soon to be swept aside by Newton's *Principia*, Descartes' redefinition of man's place in the universe had struck deeply and stuck. Descartes, so it seems, was underneath the comments of Bentley as much as Copernicus was.

In 1686, Descartes' philosophy enjoyed perhaps its most literate and widely read expression, in the classic *Conversations on the Plurality of Worlds* by Bernard Le Bovier de Fontenelle. Fontenelle—writer, philosopher, secretary of the French Academy of Sciences for half a century, important figure in the French Enlightenment—was unsurpassed in his ability to convey science to the general public. In *Conversations*, Fontenelle meets a cultured lady for several evenings of pleasant conversation. As they stroll through the park, he unfolds before her the new universe of Copernicus and Descartes—in nontechnical, witty, and poetic language. It is a universe in which nature is like a watch, and it is a universe not designed for our convenience. It is a universe in which inhabited planets orbit other suns. In Fontenelle's lifetime alone, *Conversations* went through twenty-eight editions. It was translated into English the year after its first publication and later into German. Other popular books with the same message soon appeared. As the eighteenth century got under way, the possibility of other worlds quietly slipped into Western culture.

So much for a brief history of a very big idea. Today, most of us have a modest view of our place in the universe, without thinking much about it. Our ancestors' bones look a lot like the bones of apes. We've seen pictures of our fragile planet taken from the moon. Rocks from space have crashed into our backyards.

The other day, I walked over to the house of a neighbor, a newly ordained Episcopal priest, and asked her how she felt about extraterrestrials. They were fine by her. How would she react if we made contact with them tomorrow, I asked. She'd want to know their value system, she answered.

Mirage

In southeast Persia lies the city Khashabriz. Few inhabitants have ever left its borders, for it is imprisoned within an outer city, a circle of castles and pilasters rising from the horizon like a mountain range. At times, aqueducts and windows glitter in the distance, but then dissolve. Some have ventured toward that outer fortress, only to find the castles receding in step, discouraging further exploration. It is said that, in time, a resident of Khashabriz grows resigned to confinement, walking the same cobblestone streets, passing the same food stalls filled with dates and wheat and sugar beets, breathing the same dusty air, marrying his children to the children of neighbors. When caravans and nomads sometimes drift into the city, they remain.

Like Zarathustra, the city has slowly wrapped around itself, content with isolation. No cotton in the outside world could be as silky as the cotton in Khashabriz, no pottery as delicate, no poets as enchanting. Indeed, what reason could there be to leave?

Over the years, various theories have developed among the citizens of Khashabriz regarding the origin of the distant, misty towers. One theory holds they were built by the ancient founders, to give protection from the unknown world beyond. Another claims they were erected as a blockade by foreign artisans, fearing competition with the curious silverwork and stunning carpets made within the inner city. The number of theories equals the number of people who discuss them idly in the vaulted alleys of the bazaars and on the terraces in late afternoon. On one point there is agreement: None dwell in that outer fortress because at night, while in Khashabriz the taverns and the houses glow with light, out there it is as black as coal. Except in sleep. Long ago it was discovered that the towers loom in every dream of every citizen of Khashabriz, just as in daylight they hover in the background beyond every shop, every house, every arcade.

A small group of local scientists, known for their detachment, have proposed that the surrounding castles are simply a mirage, that the people could escape at any time. They say that irregularities in the atmosphere cause light rays to bend, that the air can act as a misshapen lens, distorting some images and creating others. Similar effects, they say, disjoint the image of a spoon half in air and half in water. Most of their theorizing takes place in a little café after the evening meal and would occupy the whole of the night if their families did not call them home to sleep.

Their theory hinges on one peculiar fact: If the air density decreases with height above the ground, as happens when the temperature increases, light will bend down along its path and images will shift upward. An observer, re-creating reality by extrapolating from the light rays striking his eyes, has the impression of being inside a large bowl and sees the image of the ground curving up into a distant wall. What's more, elaborate layerings of the atmosphere can fashion turrets where there was smoothness, stripes where there was solid gray.

Few believe this explanation. Why should the air temperature day after day increase with height above the ground? The scientists answer that the land around Khashabriz by chance is cooled with a subterranean lake leading to the Gulf of Oman, while the air several meters up is warmed by constant sun and mountain breezes. Caught between cold below and heat above, the air has little choice. Why should the distant castles shimmer, as if reflecting light? The scientists answer that wind is constantly stirring the air, mingling its different densities and rapidly changing its focus. There is a final question that silences the physicists. Why have they remained in Khashabriz, if the outer fortress is just illusion? They have no answer and return to their equations, just as the baker, after listening to these strange ideas, returns to his shop.

Some of the scientists have quietly abandoned their unpopular theory, without proof or disproof. Others have become philosophers, arguing that nothing exists, that all is mirage.

With their converts they sit each day in the baths, the chambers of progressively warmer water, and do not notice whether their eyes are closed or open.

It is difficult for a stranger to understand the city Khashabriz. In some respects it is a normal city. Children run across the tiled courtyards chasing goats and sheep, lovers clench in darkened corners, morning shatters with the mullah's call to prayer. But in the middle of the night the empty streets are filled with sleepers' moanings, and on waking none can look directly at another, as if each person owed the other money. And the distant towers hover in the background, mixed with stone and air, daunting, mute.

To Cleave an Atom

In the spring of 1962 our family built a fallout shelter in the backyard. The president of the United States had been coming on the television set, pointing his finger at us, and telling us to go out and build a shelter. Some months earlier the government had distributed twenty-five million copies of a booklet called *Fallout Protection: What to Know and Do About Nuclear Attack*. I was fourteen and terrified that I would not live to be fifteen, and it was my pleading each night at the dinner table, as my three younger brothers sat quietly, that convinced my parents to dig up the backyard and put in a bomb shelter. It cost $3,000, exactly the price of the "H-Bomb Hideaway" featured in *Life* in 1955. The thing was finished just in time for the Cuban missile crisis.

A short-legged man who loved hiking created the first man-made nuclear chain reaction, on December 2, 1942, in a dis-used squash court at the University of Chicago. His name was Enrico Fermi. In Fermi's chain reaction, a subatomic particle called a neutron hits the nucleus of a uranium atom, cleaving it in two and releasing energy in the process. A uranium nucleus has quite a few neutrons of its own, and, after the split, a few of these go flying off individually along with the two main fission fragments. Each of the spawned neutrons eventu-ally strikes a fresh uranium nucleus, splitting it in half, releas-ing more energy and more neutrons, and the activity rapidly multiplies, going faster and faster. The uranium nuclei are like a lot of cocked mousetraps on the floor, each loaded with sev-eral Ping-Pong balls waiting to jump into the air when the spring is triggered. Toss a single ball into the middle to get the thing started, and soon Ping-Pong balls will be zinging everywhere. Fermi kept his chain reaction from getting out of hand by constantly removing some of the neutrons, just as the frantic release of the mousetraps can be slowed by catching some of the balls in midair before they land on cocked traps. Fermi was almost unique in twentieth-century physics for being superb in both theory and experiment. He had, with others, conceived of nuclear chain reactions in early 1939. The whole idea of fission was only a few months old at the time.

Before 1938, everyone believed that atomic nuclei remained more or less whole, with the nuclei of some elements gradu-ally disintegrating, a few small bits at a time. The emission of these bits is called radioactivity. Antoine-Henri Becquerel, a

French physicist, first discovered radioactivity from uranium in 1896, and, soon after, the husband-and-wife team of Pierre and Marie Curie observed it from another element, radium, which lost weight little by little as it hurled out tiny particles.

In the early 1900s, scientists didn't know where in the atom radioactivity originated. Atoms were pictured as solid spheres of evenly distributed positive electrical charges, embedded with negatively charged particles called electrons. The electron, discovered in 1897, was clearly a subatomic particle. Its existence already contradicted the old Greek notion that the atom was indivisible. But the details of an atom's innards were largely unknown. Then, in a brilliantly straightforward experiment in 1911, Ernest Rutherford discovered the atomic nucleus. Rutherford fired subatomic particles at a sheet of gold. The projectiles he used were alpha particles, found by the Curies in their studies of radioactivity and known to be about one-fiftieth the weight of gold atoms. If the positive charge in an atom were thinly scattered throughout its volume, as believed, then the alpha particles should have met little resistance in passing through the target gold atoms. But some bounced straight back, apparently having struck something highly concentrated. What Rutherford had discovered was that the atom is mostly empty space, with a very tiny center of positive charge, about which the electrons orbit at great distance. The dense center, the nucleus of the atom, contains all of the atom's positive charge and more than 99.9 percent of its weight. It is roughly a hundred thousand times smaller than the atom as a whole. The booming-voiced Lord Rutherford strongly preferred simple, rough-and-ready experiments, and this was surely one. He also had an excellent nose for making predictions. His experiments had shown that the atom's positively charged particles, called protons, reside in the central nucleus. Rutherford went on to predict correctly that protons share their nuclear living quarters with other, uncharged particles, later called neutrons.

One of Rutherford's collaborators from 1901 to 1903 was a man named Frederick Soddy, who later won the Nobel Prize in chemistry. They worked together on radioactivity. Soddy was impressed by the energy emerging from the depths of the atom. As early as 1903, he commented in the *Times Literary Supplement* on the latent internal energy of the atom and, in 1906, wrote elsewhere that there must be peaceful benefits for society, given the key to "unlock this great store of energy." Soddy had unusual foresight. So did H. G. Wells, who stayed well abreast of scientific developments and paid close attention to the remarks of such men as Soddy. Wells, however, made darker forecasts. In 1914 he published a lesser-known novel, *The World Set Free*, describing a world war in the 1950s in which each of the world's great cities are destroyed by a few "atomic bombs" the size of beach balls.

In many ways, the discovery of nuclear fission got under way in 1934. That was the year that Irène Curie, daughter of Marie and Pierre, and her husband, Frédéric Joliot, discovered "artificial" radioactivity. Before then, all radioactive substances had been gathered from minerals and ores. Joliot and Curie found they could *create* radioactive elements by bombarding nonradioactive ones with alpha particles. Apparently, certain stable atomic nuclei, content to sit quietly forever, could be rendered unstable if they were obliged to swallow additional subatomic particles. The forcibly engorged atomic nuclei, in an agitated state, began spewing out little pieces of themselves, just as in "natural" radioactivity. Enrico Fermi, then working in Rome, immediately took his lead from the Joliot-Curie work but decided to see if neutrons rather than alpha particles could be used to produce radioactive nuclei. Alpha particles are positively charged and therefore partly repelled by the positively charged nucleus, but the uncharged neutrons, Fermi reasoned, would have an easier time making their way into the nucleus. When these experiments proved successful, Fermi bombarded the massive uranium nucleus, containing over two hundred neutrons and protons, to see what would happen. He automatically assumed, as did others, that neutron bombardment of uranium would create nuclei close in weight to uranium. Then, in late 1938, the meticulous radiochemists Otto Hahn and Fritz Strassmann found in the remnants of bombarded uranium some barium—an element that weighs about half as much as uranium. There had been no barium in their sample to begin with. Apparently some uranium nuclei had been cut in two.

In December 1938, Hahn sent a letter describing his curious results to Lise Meitner, his coworker of thirty years. Meitner had been a respected and much loved physicist at the Kaiser Wilhelm Institute in Germany, but she was Jewish and had fled to Sweden five months earlier. At Christmas her nephew, physicist Otto Frisch, happened to pay her a visit and described the encounter: "There, in a small hotel in Kungälv near Göteborg, I found her at breakfast brooding over a letter from Hahn. I was skeptical about the contents—that barium was formed from uranium by neutrons—but she kept on with it. We walked up and down in the snow."

During their walk, Frisch and his aunt puzzled over how a single, slowly moving neutron could split in half an enormous uranium nucleus. It was well known that the protons and neutrons in an atomic nucleus are held together by strongly attractive forces—otherwise, the electrical repulsion of the protons for each other would send them flying away. How could so many attractive bonds be broken by a single neutron? Frisch and Meitner realized that the answer lay in an idea put forth by the master Danish physicist Niels Bohr. In 1936, Bohr had suggested that the particles in an atomic nucleus behave in a collective way, analogously to a drop of liquid. Frisch and Meitner reasoned that if such a drop could be slightly deformed from a spherical shape, the repulsive forces of the protons would begin to win out over the other, attractive forces. The attractive nuclear force between two nuclear particles weakens very rapidly as their separation increases, while the repulsive electrical force weakens far more slowly. Flatten a sphere of particles and each particle, on average, gets further away from its neighbors. Flatten it enough and the repulsive forces dominate, splitting it in two and sending the two halves flying apart at great speed. Frisch and Meitner calculated that the uranium nucleus was very fragile in terms of these deformations and that a small kick from a diminutive neutron might send it over the brink. According to their figures, the energy release should be enormous. Frisch went back to Copenhagen a few days later and barely managed to get the news to Bohr as the latter was boarding the Swedish-American liner MS *Drottningholm* for New York. The soft-spoken Bohr instantly slapped his head and said, "Oh, what fools we have been!" In describing the process, Frisch coined the word *fission*, by analogy with cell division in biology.

It remained for three groups of physicists, including Leo Szilard at Columbia and Walter Zinn, to demonstrate in March 1939 that neutron fission of a uranium nucleus shakes loose several new neutrons. This proved that chain reactions were possible, as Fermi had conjectured. It remained for Bohr at Princeton to calculate that only a rare form of uranium called U-235, making up about one percent of the element in nature, could sustain a chain reaction. That was why the world hadn't already blown up on its own. To build a chain reactor, U-235 had to be culled and concentrated. It could be done. It could be done by the Germans. On August 2, 1939, Albert Einstein sent a letter to President Roosevelt: "Sir: Some recent work by E. Fermi and L. Szilard . . . leads me to expect that the element uranium may be turned into a new and important source of energy in the immediate future . . . and it is conceivable . . . that extremely powerful bombs of a new type may thus be constructed. . . ."

Powerful, yes. Fissioning a gram of uranium will produce

about ten million times the energy as burning a gram of coal and air or detonating a gram of TNT. Why is nuclear energy so much more potent than any form of energy known before? TNT explosions and coal-burning release chemical energy, which has been harnessed by people in one form or another for thousands of years. Chemical energy derives from rearranging the electrons in the outer parts of atoms. Nuclear energy, of the kind we've been discussing, derives from rearranging the protons in the nucleus of the atom. Because protons are confined to a much smaller volume than electrons, their electrical "springs" are much more compressed and thus much more violent upon release. Roughly speaking, nuclear energy is more powerful than chemical energy by the same factor as the atom is larger than its nucleus. (An even more powerful form of nuclear energy works by fusing small nuclei rather than fissioning large ones.)

As Soddy predicted, nuclear energy has indeed been used for peaceful purposes. The first atomic-power generating plant began operation in Lemont, Illinois, in 1956. Unfortunately, nuclear power, which initially promised to be "too cheap to meter," has not yet proven its mettle economically. In 1984, the eighty-two nuclear plants licensed in the United States supplied only about thirteen percent of our total electric power needs and suffered from problems with management and design. Some countries in Europe have done better, but coal and oil are still the principal workhorses of the twentieth century.

What nuclear energy has dramatically changed is the meaning of war. Each new weapon in its time seemed a giant advance over its predecessors—the Roman catapult, the medieval English longbow, gunpowder artillery in the fourteenth century, TNT in 1890—but these strides were Lilliputian by comparison to the leap from chemical to nuclear weapons. Ninety-seven out of 101 of the V-1 buzz bombs aimed at London on August 28, 1944 were intercepted—a remarkable success in defense. Had these been nuclear bombs, the four that landed, in fact just one of the four, could have annihilated the whole of the city. The United States and the former Soviet Union today each possess twenty thousand such bombs, which can be launched on short notice. In our nuclear age, those ancient words of war, *defense* and *victory*, have suddenly lost their meaning. Nuclear weapons demand that we find new concepts for war and peace and weapons themselves.

Even in peacetime, nuclear weapons have violated our sense of security. In a 1980 national survey of high school students, done by Educators for Social Responsibility, eighty percent thought there would be a nuclear war in the next twenty years, and ninety percent of these felt the world would not survive it. How does one measure the psychological effects of these visions?

There is of late a widespread perception that technology, and nuclear technology in particular, has gained a momentum of its own and is hurling the world toward destruction. According to this belief, we humans are mere bystanders, helplessly awaiting our fate. I believe that our apparent helplessness regarding nuclear weapons originates from the *abstractness* of the danger more than our inability to stop it. After the destruction of Pompeii in A.D. 79, Mount Vesuvius exploded nine more times before another major eruption in 1631 destroyed many villages on its slopes and killed three thousand people. For six months prior, earthquakes shook the villages. Why did people continue to go about their business next to a working volcano? About seven hundred people were killed in the great San Francisco earthquake of 1906, and experts expect the area is due for another big one. Why do people continue to build their houses on the San Andreas fault? In these examples, as in nuclear war, the disaster has an all-or-nothing character, and its likelihood seems either small or incalculable. Of course, we cannot simply remove ourselves from nuclear weapons as we can from volcanoes and geological faults, but the psychology may be the same. Evidently, even with a choice to do otherwise, people will live in a dangerous situation, as long as the danger can be abstracted away.

The discovery of nuclear fission has gotten the world profoundly stuck, to use Freeman Dyson's word. Stuck in a buildup of nuclear weapons, stuck in outdated concepts of war and peace, stuck in human nature. If we can get ourselves unstuck, a thousand years from now people may well remember this era not so much for opening up the atom as for opening up ourselves.

Elapsed Expectations

THE LIMBER YEARS FOR SCIENTISTS, as for athletes, generally come at a young age. Isaac Newton was in his early twenties when he discovered the law of gravity, Albert Einstein was twenty-six when he formulated special relativity, and James Clerk Maxwell had polished off electromagnetic theory and retired to the country by thirty-five. When I hit thirty-five myself, I went through the unpleasant but irresistible exercise of summing up my career in physics. By this age, or another few years, the most creative achievements are finished and visible. You've either got the stuff and used it or you haven't.

In my own case, as with the majority of my colleagues, I concluded that my work was respectable but not brilliant. Very well. Unfortunately, I now have to decide what to do with the rest of my life. My thirty-five-year-old friends who are attorneys and physicians and businessmen are still climbing toward their peaks, perhaps fifteen years up the road, and are blissfully uncertain of how high they'll reach. It is an awful thing, at such an age, to fully grasp one's limitations.

Why do scientists peak sooner than most other professionals? No one knows for sure. I suspect it has something to do with the single focus and detachment of the subject. A handiness for visualizing in six dimensions or for abstracting the motion of a pendulum favors a nimble mind but apparently has little to do with anything else. In contrast, the arts and humanities require experience with life, experience that accumulates and deepens with age. In science, you're ultimately trying to connect with the clean logic of mathematics and the physical world; in the humanities, with people. Even within science itself, a telling trend is evident. Progressing from the more pure and self-contained of sciences to the less tidy, the seminal contributions spring forth later and later in life. The average age of election to England's Royal Society is lowest in mathematics. In physics, the average age at which Nobel Prize winners do their prize-winning work is thirty-six; in chemistry it is thirty-nine, and so on.

Another factor is the enormous pressure to take on administrative and advisory tasks, descending on you in your midthirties and leaving time for little else. Such pressures also occur in other professions, of course, but it seems to me they arrive sooner in a discipline where talent flowers in relative youth. Although the politics of science demands its own brand of talent, the ultimate source of approval—and invitation to supervise—is your personal contribution to the subject itself. As in so many other professions, the administrative and political plums conferred in recognition of past achievements can crush future ones. These plums may be politely refused, but perhaps the temptation to accept beckons more strongly when you're not constantly galloping off into new research.

Some of my colleagues brood as I do over this passage, many are oblivious to it, and many sail happily ahead into administration and teaching, without looking back. Service on national advisory panels, for example, benefits the professional community and nation at large, allowing senior scientists to share with society their technical knowledge. Writing textbooks can be satisfying and provides the soil that allows new ideas to take root. Most people also try to keep their hands in research, in some form or another. A favorite way is to gradually surround oneself with a large group of disciples, nourishing the imaginative youngsters with wisdom and perhaps enjoying the authority. Scientists with charisma and leadership contribute a great deal in this manner. Another, more subtle tactic is to hold on to the reins, single-handedly, but find thinner and thinner horses to ride. (This can easily be done by narrowing one's field in order to remain "the world's expert.") Or simply plow ahead with research as in earlier years, aware or not that the light has dimmed. The one percent of scientists who have truly illuminated their subject can continue in this manner, to good effect, well beyond their prime.

For me, none of these activities offers an agreeable way

out. I hold no illusions about my own achievements in science, but I've had my moments, and I know what it feels like to unravel a mystery no one has understood before, sitting alone at my desk with only pencil and paper and wondering how it happened. That magic cannot be replaced. When I directed an astrophysics conference one summer and realized that most of the exciting research was being reported by ambitious young people in their mid-twenties, waving their calculations and ideas in the air and scarcely slowing down to acknowledge their predecessors, I would have instantly traded my position for theirs. It is the creative element of my profession, not the exposition or administration, that sets me on fire. In this regard, I side with the great mathematician G. H. Hardy, who wrote (at age sixty-three) that "the function of a mathematician is to do something, to prove new theorems, to add to mathematics, and not to talk about what he or other mathematicians have done."

In childhood, I used to lie in bed at night and fantasize about different things I might do with my life, whether I would be this or that, and what was so delicious was the limitless potential, the years shimmering ahead in unpredictability. It is the loss of that I grieve. In a way, I have gotten an unwanted glimpse of my mortality. The private discoveries of new territory are not as frequent now. Knowing this, I might make myself useful in other ways. But another thirty-five years of supervising students, serving on committees, reviewing others' work, is somehow too social. Inevitably, we must all reach our personal limits in whatever professions we choose. In science, this happens at an unreasonably young age, with a lot of life remaining. Some of my older colleagues, having passed through this soul-searching period themselves, tell me I'll get over it in time. I wonder how. None of my fragile childhood dreams, my parents' ambitious encouragement, my education at all the best schools, prepared me for this early seniority, this stiffening at thirty-five.

A Visit by Mr. Newton

ONE DAY LAST WEEK I was sitting in my office at the Center for Astrophysics in Cambridge, tossing another irrelevant calculation into the wastebasket and praying to the Muses for some new ideas, when Isaac Newton walked in. I recognized him immediately from the pictures.

"How did you find the place?" I asked, a bit startled.

"Someone told me you were just south of the Holiday Inn on Mass Ave." Newton sat down, businesslike, in my extra chair. "Now, what can I do for you? My time is valuable. I've got a great little refresher course in optics, but it'll cost you."

I was in a complaining mood. "You guys had it easy. I can tell you, it's a lot harder to do scientific research these days. For one thing, every good idea I have, somebody's already thought of it. And the funding is in terrible shape. You can hardly get the necessary equipment. I wanted to buy my secretary a little word processor for typing and revising manuscripts, but the National Science Foundation won't fork over a cent.

"And who has time for research, anyway, when you have to keep current on all this," I groaned, pointing to stacks and stacks of unread journals piled up on my desk, on the floor, on the windowsill. "It must have been a joy to do science with less happening."

"What you need is another plague," Newton suggested. "I did some of my best work in 1665 and 1666, when the university was closed down and everybody was out sick."

Looking bored, Newton began snooping around my office, hesitated in front of the two stuffed and lacquered frogs playing dominoes that I had picked up in Acapulco, and finally settled in front of the bookcase. He started thumbing through the textbook on calculus and analytical geometry. "Damn. I thought I'd fixed Leibniz's wagon. I see he's still getting equal credit."

The telephone rang. It was Gruenwald at the University of Minnesota, whom I'd been trying to reach for days.

"There's another problem," I said, hanging up the phone. "How can you keep abreast of current developments when people never return your calls?"

I noticed that Newton was peering over some scribbled equations almost buried on the edge of my desk.

"What's this?" he asked.

"Oh, that. I'm investigating the electromagnetic radiation produced by a thin square sheet of gas in hyperbolic orbit around a neutron star."

"I see," said Newton, removing the sleeve of his robe from my coffee cup. "And what natural phenomena does your investigation explain?"

"Uh, it's a theoretical problem, of course. But my calculations should be a perfect test of the Ludwick-Friebald effect," I answered shrewdly. "A graduate student in Cincinnati is extremely interested in the result."

"Trifling. Hath not anything important in science transpired since my *Principia*?"

Newton was strict, but I would try to impress him. "Let's see. Darwin showed that species evolve by survival of the fittest. Einstein discovered that the flow of time is relative to the observer. De Broglie and Heisenberg and Schrödinger found that particles actually behave as waves and can be in several places at the same time. Watson and Crick discovered the structure containing the blueprint for reproducing life. We've developed very fast devices for mathematical computations, called computers, which are gradually taking over society. And a while back men landed on the moon."

"Cheese?"

"No, sorry. I'm afraid that particular theory of yours didn't pan out. Oh, I almost forgot. A few years ago some guys invented a perpetual motion machine called supply-side economics."

Newton had that impatient look again, and I realized time was running out on a precious opportunity. All I wanted were a few profound ideas. To be honest, the Ludwick-Friebald effect was beginning to bore me. I'd just love to prance into the next meeting of the American Physical Society with some brilliant new equations and show them I had the right stuff, after all.

Newton was flipping through my books again—differential equations, thermodynamics, quantum mechanics, radiation theory—muttering "trivial" after each book.

"Look, Mr. Newton," I said. "Here's a pad of paper and a pencil. I'd be grateful if you would write down some original results for me. Maybe something like a fourth law of motion. Or perhaps a new theory of elasticity."

Newton sat down at my desk, shoving aside the last ten issues of the *Astrophysical Journal*, and was silent for several minutes. "Well, I do have one item for you. I hate to admit it," he said sheepishly, "but I made a mistake in my universal law of gravitation. The force of gravity varies as the inverse cube of the distance, not the inverse square."

"You're kidding."

"No. I had to get it off my chest. You're the first person I've told."

This was big. In fact, this was so big maybe I should keep it secret for a while and milk it for all it was worth before announcing the news to my colleagues. Fernsworth at Princeton would be green with envy. Eventually NASA would certainly need to know. And the Pentagon, before the Russians found out. Come to think of it, this stuff would probably be classified the moment it leaked out.

"How did you realize you'd made a mistake?" I asked, after calming down.

"I could never account for this vicious left hook in my long wood shots, despite my best calculations and attempts to compensate. Finally, I decided the fundamentals had to be wrong, so I worked the problem backward and deduced the inverse cube law. I was too embarrassed to tell anybody at the time."

I gave a long, low whistle. This was even bigger than I thought. Palmer, Nicklaus. They could probably shave five strokes if they knew what I now knew. The implications of the inverse cube law were staggering.

"I want you to know," said Newton, "that I take no responsibility for the consequences. *Hypothesis non fingo*."

"I understand." My mind was racing. Many things were beginning to fall into place. Mysterious phenomena that had always puzzled me now made perfect sense. Suddenly I could explain why Aunt Bertha always had trouble getting up from the dinner table. And why my folded pants always slid off the bedroom chair during the night. The more I thought about it, the logic of the inverse cube law seemed inescapable.

Newton was leaning back in his chair, exhausted. His eyes were glazed. I found myself warming up to the old boy since he had humbled himself in front of me. For the next half hour we talked of lighter subjects—some optics, a bit of kinetics, a little alchemy. Then he rose and, reciting a few lines from *Paradise Lost*, walked out of my office.

It has been a week now since Mr. Newton's visit. For the first few days I was paralyzed with the knowledge of the inverse cube law. Antigravity. Ketchup that pours instantly. New weapons of destruction. I could not sleep, I could not work.

Finally, I pulled myself together and nervously began to calculate. I bungled the first equation, wadded up the paper and lobbed it toward the garbage can, and began again. Out of the corner of my eye I noticed that the paper wad bounced off the blackboard, skidded along a file cabinet, knocked over one of the frogs from Acapulco, and landed cleanly in the can. Curious. That trajectory was exactly as predicted by Newton's original theory of gravity. I put down my pencil and threw another wad, then another, then another. Every shot confirmed the old inverse square law. I tipped over stacks of journals and clocked their motion, leaped repeatedly off my desk, hurled books across the room. Slowly, the stubborn truth began to dawn on me; Mr. Newton's original theory had always been right. I guess after all those centuries Newton had grown senile. They do say physicists peak at an early age.

Things have settled down, although my office is a mess. After a short vacation, I'm going back to the Ludwick-Friebald effect. It may not be important, but it's probably correct.

ORIGINS

WHEN I WAS A CHILD and asked my parents where I came from, they referred me to the copy of *The Stork Didn't Bring You* nestled in the den library, and that was that. While still in awe of the biological details, I became a physicist. Physicists, who by profession think in the simplest terms possible, have their own version of the story. Brushing aside questions of egg and cell development, evolution of the species, and so on, they get down to atoms in the body. A physicist's answer to where we came from is an investigation of the origins of the chemical elements. As it turns out, we were all made in stars, some five to ten billion years ago.

All living matter we know about is composed mainly of hydrogen, carbon, oxygen, nitrogen, phosphorus, and sulfur. Carbon, with its rich variety of chemical bonds, is particularly suited for forming the complex molecules that life thrives on. Atoms of all these elements are continuously cycled and recycled over many generations through our planet's biosphere, incorporated into plants from the soil, swallowed by animals, inhaled and exhaled, evaporated from oceans, and returned to soil, air, and sea. We've been trading atoms with other living things since life began.

But where did the atoms come from? One rather bland possibility is that they were always here, in their observed proportions, thus putting a stop to further delicate questions. A great deal of scientific evidence, however, suggests this is not the case. First of all, the Earth is radioactive. Atoms of various elements are constantly aging and changing into other atoms by the ejection and transformation of protons and neutrons in their nuclei. For example, uranium 238, consisting of 92 protons and 146 neutrons, changes into thorium 234 by the simultaneous emission of two protons and two neutrons. Thorium is itself unstable and decays into another element and then another, until lead 206 is produced. Lead, at last, is mature, and the transformation process comes to a halt.

For years, chemists and physicists have been taking census reports of these busy families of atoms, with their noisy infants, teenagers, and quiet senior citizens. It seems perfectly natural that the relative proportions of the elements had to be different in the past. How far in the past? Analyses of the observed numbers of uranium versus lead atoms, for example, have determined that the Earth is 4.5 billion years old. Our atomic roots go back this far and further.

The best guide to what was happening so long ago is found in the vast reaches of space, beyond our solar system, beyond our galaxy of one hundred billion stars, beyond our neighboring galaxies. When we peer out through our telescopes at distant galaxies, hundreds of millions of light-years away, we find them receding from us. The universe is and has been in a state of expansion, with the galaxies rushing away from each other like painted dots on an expanding balloon. Running this scene backward in time suggests the universe began about ten billion years ago, in an initial explosion called the Big Bang. At this point even the people who work on these things start getting wide-eyed, despite the logic of equations, computers, and telescopes.

The early universe was much denser than today's. And it was much hotter, just as squeezing an ordinary gas tends to raise its temperature. When the universe was sufficiently young and hot, none of the chemical elements except hydrogen 1 (whose nuclei are single protons) could have existed. The constituent protons and neutrons of any compound atomic nucleus would have simply evaporated under the intense heat. For example, carbon and nitrogen atoms would have disintegrated into unattached protons and neutrons at temperatures exceeding about 2,000 billion degrees Fahrenheit. According to cosmological theory, this was the case until a tenthousandth of a second after the Big Bang. As far as we can tell, the infant universe held only a shapeless gas of subatomic particles. Atoms, stars, planets, and people came later.

With only some introductory thermodynamics, a little cosmology, and some whisperings from nuclear physics, we have narrowed down the origin of the elements to sometime after the universe began but before the formation of the Earth. Where and when did this occur?

Results in nuclear physics indicate that, starting from a hot gas of unattached protons and neutrons, synthesis of complex atoms proceeds along a family tree, in which heavier atoms grow from lighter ones. Since the temperature and density of the expanding, primeval universe were dropping rapidly with time, there was only a brief period, ending a few minutes after the Big Bang, when conditions were right for creating elements. Before this period, every partnership of two or more particles evaporated; after this period, the subatomic particles did not have the energy and were too far apart for fusion to occur easily. According to theoretical calculations, element formation in this delicate interval got only as far along as helium 4 (two neutrons and two protons), the lightest element after hydrogen. The predicted amount of helium produced, about twenty-five percent of the mass in initial protons and neutrons, is in delightful agreement with current-day observations of the cosmic helium abundance. Nice, but what about carbon, oxygen, and other elements?

The answer, as we now understand it, began emerging in 1920 when the eminent British astronomer Sir Arthur Eddington first proposed that the sun and other stars are powered by nuclear fusion reactions. This is the same source of energy that is unleashed, for ghastly purposes, in our hydrogen bombs. In the deep interior of stars, densities and temperatures can become sufficiently high to fuse lighter elements into heavier ones, going far beyond helium. Such observed features of stars as their masses, temperatures, and luminosities accord well with the theoretical models and provide indirect confirmation of the hypothesized nuclear reactions. These are the facts of life that adult physicists and astronomers will tell you.

More direct evidence for the element-producing activity of stars comes from analysis of the debris ejected by exploding stars. In such explosions, called supernovae, nuclear reactions proceed at an extremely rapid rate; both the hastily produced elements and those manufactured in the preceding, more leisurely evolution of the star are spewed out into space, where we can have a good look at them. Analyzing the telltale colors of light emitted by stellar ejecta reveals a host of heavy elements, in the relative proportions predicted by nuclear physicists.

The first stars could have begun forming long ago, when the universe was only a million years old. In fact, we see evidence for a great spread in the ages of stars. New stars are continually being born. Relatively young stars, like our sun, and its surrounding planetary system have condensed out of gas enriched with the drifting fragments of ancestral stars, gas thus enriched with heavy elements.

As we go about our daily business on this small planet, we have little feeling for the bond between us and those distant points of light. Excepting hydrogen and helium, all the atoms in us and our biosphere were bred somewhere in space, in the nuclear reactions of some now defunct star.

A Day in December

Shortly before six o'clock on Thursday morning, December 6, 1979, someone's dog, let loose, ran yelping down Embarcadero Road in Palo Alto, turned right on Waverley, and dropped from fatigue and boredom near the intersection of Santa Rita Avenue, having woken all sleepers within earshot. It was still dark. Lights blinked on one by one along the animal's path, people groped for robes and went to the toilet, and another day began.

By half past seven University Avenue was filling with college students cycling to their early classes. At the doorway of one of the large homes on Waverley near University, a woman in her early forties, wearing a smart tweed suit, called out to her husband, "George, don't forget the gardening book for Betty." George, in pinstripes, nodded and drove away to a Silicon Valley company.

A couple of hours later, in his rented house on Camino a Los Cerros, Alan Guth got out of bed, had two hard-boiled eggs, and waved good-bye to his wife and son (who the day before had said, "Daddy's home," for the first time). On his ten-speed, which was kept in good repair with supplies from the Palo Alto Bike Shop, he rode southeast to Sharon Road, turned right, flew past the shopping center, turned left onto Sharon Park Drive, turned right on Sand Hill Road, and entered the grounds of the Stanford Linear Accelerator Center. His office was on the northeast corner of the third floor of Central Laboratory, in the theory group. Guth was a thirty-two-year-old physicist.

By this time it was half past ten. Students and student-types were beginning to hang around Printer's Inc., a bookstore on California Avenue with a coffee bar and classical music in the background. A well-fleshed man in corduroys was thumbing through *Diet for a Small Planet*, wondering what to serve at his vegetarian dinner party.

On the street outside, the day was fine, unseasonably fine. The woman in the smart tweed suit, on her way to look at new wallpaper, decided to go home and change into something cooler. The weatherman had predicted rain. She hurried. Her old wallpaper of seven years, brimming with five-inch bur-gundy squares caught within a thicket of yellow diagonal stripes, had to go.

Guth started work with coffee. His colleagues on the third floor shared a community coffeepot for $3 a month per person. Around noon, after placing an anxious phone call about a possible job for the coming year, Guth went with two friends to lunch at the New Leaf. Afterward, back in his office, he wrote some correspondence—he did all his writing with a Radiograph pen, with its bold, neat lines—and later discussed magnetic monopoles and cosmology with a colleague. At six o'clock Guth pedaled home. Cedar, Camino de Los Robles, Monterey, Manzanita, Camino a Los Cerros. He knew the side streets on his route. In fifteen minutes he was home, had a broiled steak, medium rare, and after dinner he and his wife did laundry. He was out of underwear.

The man in corduroys with the dinner party outdid himself. At evening's end, he slumped exhausted on his sofa, leaving the dirty plates and enameled soufflé pans in stacks around the kitchen. Half an hour of television before bed would smooth him out. Click. A floor wax commercial.

Outside, above Palo Alto, the deep blue sky was blackening. Higher still, uncounted stars cut silently into the night. Sometime between eleven and twelve o'clock, sitting at his study desk with only pen and paper, Guth discovered mathematical evidence that, contrary to previous theories, the infant universe ten billion years ago underwent a fantastically rapid expansion, just after which the matter that was to form atoms and galaxies and people came into being.

Progress

OVER THE PAST SEVERAL YEARS, friends and colleagues have become increasingly irritated with me for not being on the electronic network. Scientists want to send me their data on E-mail. Secretaries for distant committees, forced to resort to the telephone, hound me for my E-mail address and lapse into stunned silence when I allow that I don't have one. University administrators, who organize meetings and send messages across campus at the push of a button, grumble about hand-carrying information to me or, even worse, putting paper in an envelope and sending it through the interdepartmental-mail system. I admit I'm a nuisance. But I resist getting on the Internet as a matter of principle, as a last holdout against the onslaught of unbridled technology galloping almost blindly into the twenty-first century.

For at least the past two hundred years, human society has operated under the assumption that all developments in science and technology constitute progress. According to that view, if a new metal alloy can increase the transmission of data from 10 million bits per second to 20 million, we should create it. If a new plastic has twice the strength-to-weight ratio as the older variety, we should produce it. If a new automobile can accelerate at twice the rate of the current model, we should build it. Whatever is technologically possible will find an application and improve us.

The ordained imperative of advancing technology was probably thrust on its course at the start of the Industrial Revolution, although the idea must have had some velocity before then. As everyone knows, new technology in the eighteenth century, such as the power loom and the steam engine, dramatically improved the efficiency of production, with the associated financial rewards. Power looms enabled textile workers to perform at ten or more times their previous rate—and the machines never grew tired. Steam engines, which could produce up to one hundred times as much power per weight as humans and oxen, transformed England into an industrial and economic giant. With such outcomes, it was natural to equate technology with progress.

But that equation expressed far more than the obvious connection between technology and material improvement. The concept of progress was a major intellectual and cultural theme of the last century, fueled not only by the Industrial Revolution but also by the new theory of evolution. Many scientists and nonscientists of that period interpreted biological evolution as a kind of progress from lower forms to higher, culminating in human beings. By extension they believed that natural (biological) and human-made (technological) forces were together causing society to become more developed, more organized, and more moral with time. Progress was part of our manifest destiny. Writers, philosophers, and social thinkers, as well as scientists and engineers, took up those more general ideas. Following the ideals of the Enlightenment, progress necessarily included social and political advances too. For example, Edward Bellamy's classic novel *Looking Backward* (1888), set in Boston, describes an ideal social and industrial system of the future.

In the twentieth century the concept of progress changed, becoming increasingly tied to technology and large dehumanized technological systems. By the time of the 1939 World's Fair, in New York, one could read the following in the promotional literature of the futuristic General Motors exhibit: "Since the beginning of civilization, transportation and communication have been keys to Man's progress, his prosperity, his happiness." In one fell swoop, technology, progress, and happiness had become bound in a compelling dream of the future.

Today, at the end of the twentieth century, a crucial question before us is whether developments in technology inevitably improve the quality of life. And if not, we must ask how our society can employ some selectivity and restraint, given the enormous capitalistic forces at work. That is a terribly difficult problem for several reasons, not the least of which is the subjective nature of progress and quality of life. Is progress greater human happiness? Greater comfort? Greater speed in personal transportation and communication? The reduction of human suffering? Longer life span? Even with a definition of progress, its measurements and technological requirements are not straightforward. If progress is human happiness, has anyone shown that twentieth-century people are happier than nineteenth-century people? If progress is comfort, how do we weigh the short-term comfort of air-conditioning against the long-term comfort of a pollution-free environment? If progress is longer life span, can we ever discontinue life support for a dying patient in pain?

Only a fool would claim that new technology rarely improves the quality of life. The electric light has expanded innumerable human activities, from reading to nighttime athletic events. Advances in medicine—particularly the germ theory of disease, public-health programs, and the development of good antiseptics—have obviously reduced physical suffering and substantially extended the healthy human life span.

But one can also argue that advances in technology do not always improve life. I will skip over such obvious environmental problems as global warming, ozone depletion, and nuclear-waste disposal, and consider something more subtle: high-speed communications. We are already seeing people at restaurants talking into cellular phones as they dine. Others take modems on vacations, so they can stay in touch with their offices at all times. Or consider E-mail, the example I began with. E-mail has undeniable benefits. It is faster than regular mail and cheaper and less obtrusive than the telephone. It can promote conversations among far-flung communities of people, and it can encourage otherwise reticent talkers to speak up, via computer terminals. But E-mail, in my view, also contributes to the haste, the thoughtlessness, and the artificial urgency that increasingly characterize our world. The daily volume of E-mail communications is inflating without limit. A lawyer friend says he spends 50 percent of his time at work sifting through unimportant E-mail messages to arrive at the few that count. Some communications are invariably of the form "Please ignore my last message." Evidently, it has become so easy and fast to communicate that we often do so without reflection. When messages come in so quickly and effortlessly, we irresistibly and immediately respond in kind. Although I cannot document it, I suspect that bad decisions are being made because of the haste of transmitting and responding to E-mail messages. But more to the point is the overall fast-food mentality at work in the rapid conveyance of our thoughts and responses. We are suffocating ourselves. We are undercutting our contemplative powers. We could even be, ironically, impeding progress.

E-mail, of course, is only one example. Its use or abuse is up to the individual. But E-mail is representative of other technological developments, such as genetic engineering, throwaway plastics, advanced life-support systems, and computer networks. Certainly, many of those developments will have good consequences. But that is not the point. Modern technology is racing forward with little examination or control. To be sure, a number of thinkers and writers have for some time expressed alarm over where unchecked science and technology might be taking us. Mary Shelley, in *Frankenstein* (1818), was certainly concerned about the ethical dilemmas of artificial life. So was H. G. Wells in *The Island of Dr. Moreau* (1896), wherein the evil surgeon, Dr. Moreau, synthesizes creatures that are half man and half beast. In *Walden* (1854), Thoreau wrote, "We do not ride on the railroad; it rides upon us." A more recent example is Don DeLillo's *White Noise* (1985), in which the hero is exposed to a cloud of poisonous industrial chemicals, and then suffers a far worse, mental ailment because of a computerized medical system that constantly announces his fate. But those countervailing voices have, for the most part, been ignored. That is not just because of the considerable economic forces that are propelling today's ravenous technological engine. Rather, we seem to believe—perhaps at some subconscious level—that technology is our sacred future.

I am not in favor of squashing new developments in pure science, in any form. The act of understanding the workings of nature—and our place in it—expresses for me what is most noble and good in us. As for the applications of science, I am certainly not opposed to technology as a whole; I benefit greatly from it. But we cannot have advances in technology without an accompanying consideration of human values and quality of life.

How should this examination and questioning proceed? I don't know. It is not likely that government regulations would be effective. Our government, as well as other large institutions, understandably has an investment in allowing technology to develop unabated. The problem cannot be solved from the top down. It is a cultural problem. Perhaps we must regulate ourselves. Perhaps we each must think about what is truly important in our lives and decide which technologies to accept and which to resist. That is a personal responsibility. In the long run, we need to change our thinking, to realize that we are not only a society of production and technology but also a society of human beings.

I = V/R

I was somewhat embarrassed not so long ago when I opened a year-old physics journal and read that two Japanese fellows had attacked the same problem I was currently finishing up, obtaining an identical solution. The problem, not so consequential now as I reflect stoically on my preempted calculations, concerned the spatial distribution that would eventually be achieved by a group of particles of different masses interacting with each other by gravity.

The underlying theories of gravity and of thermodynamics necessary for solving such a problem are certainly well established, so I suppose I should not have been surprised to find that someone else had arrived at similar results. Still, my pulse raced as I sat with my notebook and checked off each digit of their answers, in exact agreement with mine to four decimal places.

After doing science for a number of years, one has the overwhelming feeling that there exists some objective reality outside ourselves, that various discoveries are waiting fully formed, like plums to be picked. If one scientist doesn't pick a certain plum, the next one will. It is an eerie sensation.

This objective aspect of science is a pillar of strength and, at the same time, somewhat dehumanizing. The very usefulness of science is that individual accomplishments become calibrated, dry-cleaned, and standardized. Experimental results are considered valid only if they are reproducible; theoretical ideas are powerful only if they can be generalized and distilled into abstract, disembodied equations.

That there are often several different routes to a particular result is taken as an indication of the correctness of the result, rather than of the capacity for individual expression in science. And always there is the continual synthesis, the blending of successive results and ideas, in which individual contributions dissolve into the whole. Such strength is awesome and reassuring; it would be a tricky business to land a man on the moon if the spaceship's trajectory depended on the mood of the astronauts, or if the moon were always hurrying off to unknown appointments.

For these same reasons, however, science offers little comfort to anyone who aches to leave behind a personal message in his work, his own little poem or haunting sonata. Einstein is attributed with the statement that even had Newton or Leibniz never lived, the world would have had the calculus, but if Beethoven had not lived, we would never have had the C-minor Symphony.

A typical example of scientific development lies in the work of the German physicist Georg Simon Ohm (1789–1854). Ohm was no Einstein or Newton, but he did some good, solid work in the theory of electricity. Coming from a poor family, Ohm eagerly learned mathematics, physics, and philosophy from his father. Most of Ohm's important research was done in the period 1823–1827, while he worked grudgingly as a high school teacher in Cologne. Fortunately, the school had a well-equipped physics laboratory. In 1820 Hans Christian Oersted had discovered that an electric current in a wire could affect a magnetic compass needle, and this development impassioned Ohm to begin work in the subject. In those days electrical equipment was clumsy and primitive. Chemical batteries, invented in 1800 by Alessandro Volta and known as voltaic piles, were messy affairs, consisting of ten or more pairs of silver or copper and zinc disks separated by layers of moist cardboard. Ohm connected a wire to each pole of a voltaic pile and suspended above one of the wires a magnetic needle on a torsion spring. This was a crude device, operating on Oersted's principles, which could measure the current flowing through a wire. Ohm then completed the circuit by inserting test wires of various thicknesses and lengths between the two battery leads, measuring how the current changed and depended on the properties, or "resistance," of each test wire.

This initial work was done inductively, by the seat of the pants. Ohm published his results in a semiempirical form, smacking of the flavor of the laboratory. Some of the quantitative expressions of experimental data in the first paper in

1825 are actually slightly incorrect. This was soon to be rectified, however, for Ohm was enamored of the elegant and mathematical work of Jean Fourier on heat conduction and recognized some striking similarities to current flows. Under this influence, Ohm further developed and recast his results into more general mathematical expressions, not exactly matching his data but cleaving to the analogies with Fourier's work, a creative and crucial step.

The final results, stating in part that the current is directly proportional to voltage and inversely proportional to resistance (now universally known as Ohm's Law), were codified in an abstract and well-manicured paper published in 1827, very distant from those late nights with jumbles of wires and repeated exhortations to the voltaic pile to hold steady on the voltage.

When the complete theory of electromagnetism was assembled by James Clerk Maxwell in 1864, Ohm's work was deftly stitched in, like a portion of a giant tapestry. In 1900 Paul Drude published the first microscopic theory of resistance in metals, giving at last a satisfying theoretical understanding of Ohm's Law. Today, we use Ohm's Law routinely in designing electrical circuits, in calculating how deep a radio wave will penetrate into the ocean, and so on. But there is little of Ohm in the abstract statement $I = V/R$ (current equals voltage divided by resistance).

Max Delbrück, the physicist-turned-biologist, said in his Nobel Prize address, "A scientist's message is not devoid of universality, but its universality is disembodied and anonymous. While the artist's communication is linked forever with its original form, that of the scientist is modified, amplified, fused with the ideas and results of others and melts into the stream of knowledge and ideas which forms our culture." Perhaps if Georg Ohm had been a painter or a poet, we would now be celebrating his leaky voltaic pile, his uncalibrated galvanometer, his exact arrangement of odd wires and mercury bowls, or reliving the loneliness of his bachelor nights, his emotions and thoughts during the experiments.

It seems to me that in both science and art we are trying desperately to connect with something—this is how we achieve universality. In art, that something is people, their experiences and sensitivities. In science, that something is nature, the physical world and physical laws. Sometimes we dial the wrong phone number and are later found out. Ptolemy's theory of the solar system, in which the sun and planets revolve about the Earth in cycles and cycles within cycles, is imaginative, ingenious, and even beautiful—but physically wrong. Virtually unquestioned for centuries, it was ungracefully detonated like a condemned building after Copernicus came along.

Very well. Scientists will forever have to live with the fact that their product is, in the end, impersonal. But scientists want to be understood as people. Go to any of the numerous scientific conferences each year in biology or chemistry or physics, and you will see a wonderful community of people chitchatting in the hallways, holding forth delightedly at the blackboard, or loudly interrupting each other during lectures with relevant and irrelevant remarks. It can hardly be argued that such in-the-flesh gatherings are necessary for communication of scientific knowledge these days, with the asphyxiating crush of academic journals and the push-button ease of telephone calls.

The frantic attendance at scientific conferences has been referred to as a defense of scientific territoriality, a dead giveaway to our earthy construction. I think it is this and more. It is here, and not in equations, however correct, that we scientists can express our personalities to our colleagues, relish an appreciative smile, speculate on the amount of Carl Sagan's latest royalty advance, and exchange names of favorite restaurants. Sometimes I enjoy this as much as the science.

Nothing but the Truth

In a new preface to his first novel, Italo Calvino reminds us that writers mold reality to fit their purposes; landscapes are distilled, remembered faces are tortured. Art demands interpretation and recasting of the naked experiences of life. To some extent the same is true of science. Nature does not reveal herself in easy glimpses of scientific truths. Experimental results are often confusing and sometimes plain wrong. Without an interpretive theory, without a design offered by the beholder, observations of the physical world are just so many loose, meaningless facts.

Little wonder then that the history of science is replete with personal prejudices, misleading philosophical themes, players miscast. Prejudice is a dirty word in science, whose musty corridors were supposedly swept clean by Copernicus and

Galileo. Yet I suspect all scientists have been guilty of prejudice at various times in their research.

An unexpected example can be found in the work of Lev Davidovich Landau, winner of the 1962 Nobel Prize in physics. Among other things, Landau made major contributions to the theories of ferromagnetism, superfluids and superconductivity, and suggested the fundamental law of charge-parity conservation. Landau pioneered the school of modern Soviet theoretical physics and was practically worshiped by colleagues. He was also feared, partly for his habit of ruthlessly ferreting out and destroying all unproven statements in scientific discussions. The nameplate on his office door at the Ukrainian Physicotechnical Institute read: "L. Landau. Beware, he bites."

One of Landau's favorite remarks was "nonsense always remains nonsense." In 1932, with several important pieces of work already under his belt, Landau published a curious three-page paper called "On the Theory of Stars." The paper begins with high expectations, quickens through its penetrating and elegantly simple calculations, and ends in nonsense.

What is shocking about the 1932 paper is that Landau, without warning and in a single sentence, dismisses a major branch of physics. The paper concerns a theoretical investigation of the structure attained by stars in balancing their inward gravitational forces against their outward pressure forces. For the burned-out stars Landau was considering, the outward pressure forces are prescribed by quantum mechanics, the theory of matter at the atomic level. By 1932 the laws of quantum mechanics had been firmly established and ranked beside Einstein's relativity as the foundation for modern physics. To Landau's dismay, his calculations predicted that burned-out stars cannot avoid complete inward collapse if slightly more massive than the sun. That is, in sufficiently massive cold stars no amount of internal pressure can counterbalance the inward crush of gravity, leading to a frantic contraction of the star from a sphere a million miles or more across down to a point. Landau then writes: "As in reality such masses exist quietly as stars and do not show any such ridiculous tendencies, we must conclude that . . . the laws of quantum mechanics are violated." (Sir Arthur Eddington made an almost identical remark at a Royal Astronomical Society meeting in 1935, upon reviewing calculations by S. Chandrasekhar that independently reached the same conclusions about cold stars.)

Landau had little justification for this statement. Puzzling, for such a meticulous scientist. Astronomers had indeed observed very massive stars quietly avoiding collapse, but these were clearly not the burned-out, cold stars Landau's calculations applied to. Hot and cold stars were, in fact, easily distinguished by their colors. In his mistaken reference to observed stars we see a mirror turned not outward to external reality, but inward. It seems Landau found his theoretical result (which was actually one of the first predictions of black holes) so preposterous, so disturbing to common sense, that he was willing to abandon the celebrated theory that produced the result. And in the same, concise prose that normally followed logically from his calculations.

Landau's paper was not the first example of personal prejudice in science, nor the last. In 1917 Einstein modified, in a completely ad hoc manner, his 1915 theory of gravity because it predicted a dynamic universe, a cosmos always on the move, either expanding or contracting. Since Aristotle, the static nature and permanence of the universe had been simply accepted in Western thought, like night changing into day. There was certainly no observational evidence to the contrary. Einstein succumbed to this bias. Where his original equations were dispassionate, he was not. Einstein realized his mistake in 1929, when the astronomer Edwin Hubble looked at distant galaxies through a telescope and found that the universe was expanding.

Landau and Einstein might be forgiven for placing too much trust in their physical intuition. To the theoretician at the drawing board, reminiscences of physical reality are a valuable tool—an important distinction between science and mathematics. When totally unfamiliar results like black holes and expanding universes peer out from the equations, even fearless intellects sometimes retreat.

With observational science, personal bias can take a subtle form. In 1969 Joseph Weber of the University of Maryland reported positive evidence for the first detection of gravitational radiation, the weaker cousin of electromagnetic radiation and long predicted on theoretical grounds. In the subsequent decade other scientists repeated Weber's experiments with more sensitive equipment, but obtained only negative results.

In 1975 P. Buford Price of the University of California at Berkeley and collaborators announced evidence for the detection of magnetic monopoles. Magnetic monopoles, if they exist, would be the magnetic version of electrically charged particles, such as electrons, and would place electricity and magnetism on an equal footing. Many physicists have long been troubled by the apparent absence of magnetic monopoles. A theory for such particles has been in hand since 1931. But Price almost certainly misinterpreted his data, as later analyses by other scientists showed. Both Weber and Price were earnestly stalking their prey. In science, as in other activities, there is a tendency to find what we're looking for.

And it's not surprising. After a day in the laboratory with the purring instruments or the silent equations, scientists return to the world of other men and women. Listen in at a scientific conference while people are presenting the results of their research. If a scientist has given himself to the project, you'll hear more than summaries of data and procedures. Chances are you'll hear a lively commentator, an advocate of a particular point of view, a man or woman trying to make sense of things in their own terms. As Bacon shrewdly observed, "The human understanding is no dry light, but receives an infusion from the will and affections; whence proceed sciences which may be called 'sciences as one would.' For what a man had rather were true he more readily believes."

Fortunately, the scientific method, that legendary code of detachment and objectivity, does not hang on the actions of individual scientists. Instead, it draws strength from collections of scientists, experiments repeated to confirm or deny, theories considered and reconsidered by skeptics. Scientists may defend their own ideas with unseemly dedication, but they relish finding flaws in the work of colleagues. Most personal prejudices crumble under such an onslaught of devil's advocates.

There's another source of unbridled hypothesis from which science, even as a collective enterprise, is not immune. And that's the great distance between theory and experiment in some areas, leaving portions of theories adrift and stranded, without easy approach. Einstein's theory of gravity is well tested in the solar system, but it also makes critical predictions about black holes, where gravity is a million times stronger than at the sun. The debate in biology over gradual versus catastrophic change in evolution of species remains controversial partly because a decisive fossil data base is hard to come by. Scientific theories stand or come tumbling down on their predictions; when predictions outstride our ability to test them, we've entered dangerous territory. My own cautious prediction is that this will remain a problem we have to live with. Even in science, our minds can flutter to heights where bodies cannot follow.

Some years ago I heard a commencement address by Richard Feynman, a Nobel Prize winner who worked on some of the same problems as Landau. It was a warm June morning. The future scientists sat in folding chairs on the lawn, perspiring in their black gowns. But no one noticed the heat. Hundreds of young faces were fastened on the podium, where Feynman was giving advice. He said that when we do scientific research, when we publish our results, we should try to think of every possible way we could be wrong. His words hovered in the thick air, blending with the various ambitions and beliefs gathered there. It was a tall order.

TIME FOR THE STARS

MORE THAN ONCE IN THE PAST DECADE, I've been drawn into a heated discussion over the vast sums being budgeted toward a military defense in space. Much of this money has been suddenly offered to the scientific community. Should it be accepted? For me, in addition to the ethical and practical questions, this raises the issue of applied science versus pure science.

I am worried that our country has become increasingly shortsighted to the value of pure science. One recent example was the court-ordered breakup of American Telegraph and Telephone, leaving its basic research group, Bell Laboratories, in a vulnerable spot. Another was the congressional veto of a relatively cheap exploratory mission to Halley's Comet, which visits the solar system only once in a lifetime. Certainly, few people would deny the material comforts, the economic advantages, the power to make war or peace that applied science brings. But in pursuing these other goals, we have paid less and less attention to the value of science for its own sake.

Our national preoccupation with applications goes back to the cultural and political origins of our country. In Europe, science was traditionally considered a part of culture, and a person could devote his life to science as a gentleman and a scholar. Isaac Newton, as a fellow of Cambridge University, needed no justification for his studies of physics. Carl Friedrich Gauss, who made brilliant contributions to mathematics and astronomy in the early nineteenth century, was supported through the patronage of the duke of Brunswick. In contrast, when science got under way in America, in the middle 1800s, the democratic ideals of our young country demanded a direct accounting to the people, a direct benefit to society. Scientific research was usually supported only if it was part of a practical or technical enterprise, like the U.S. Weather Service, founded in 1870, the U.S. Geological Survey, founded in 1879, or the National Bureau of Standards, founded in 1901. Gradually, our nation began to take pride in and identify with its technological achievements (which exclude pure science by definition). The American hero of science is Thomas Edison, not Willard Gibbs, who made fundamental contributions to the theory of heat. During World War I, even the great physicist Robert Millikan said that "if the science men of the country are going to be of any use to her, it is now or never." Since World War II, of course, our country and all countries have been keenly aware that military might can be achieved through science.

Also since World War II, science has become a big business. In many areas of science, the romantic days of the lone scientist, uncovering the secrets of nature with homemade equipment, are gone. Experiments today often require large teams of scientists, large budgets, and large bureaucracies to manage them. Some of these operations could not be mounted unless they utilized the existing military-industrial complex. And the pace of society in general has quickened. Under constant pressures, we grasp for the short-term payoff.

Why should our nation, or any nation, support pure science? Why should a nation pay for an activity that brings it no clear economic or military advantage? Why should a nation support an activity that seems *useless*?

It seems to me that pure science has several different values. In order of increasing range into the future, pure science entertains us, it provides the soil from which technology grows, it changes our worldview, and it grants us cultural immortality.

On an immediate, day-by-day basis, learning new things pleases us, and there is no doubt that we learn from pure science. Furthermore, what we learn is "true," it concerns the real world, and it can be understood in broad terms by every intelligent person. Nonscientists are entertained by learning what comets are made of, just as they are entertained by seeing a new Neil Simon play or reading a new book by Gabriel García Márquez. Everyone is a potential consumer of pure science. If pure science cannot pay for itself in the marketplace, as movies and books do, it is perhaps because its pleasures lie in knowledge. Still, this knowledge brings a special kind of happiness, and the happiness of a nation's people counts for something.

Pure science may seem useless in the usual sense, but over a long period of time it surely leads to economic and technological benefits. If we stop paying for pure science today, there will be no applied science tomorrow. Darwin's work on evolution and Mendel's on the heredity of plants laid the foundations for the science of genetics, which eventually led to the discovery of DNA, which led to genetic engineering, which is now exploding with unimaginable applications. Faraday's discovery of how a magnet can produce electricity made possible the first hydroelectric power plant, fifty years later. Yet Darwin and Mendel and Faraday were not supported with any such profits in mind, nor could they have been. A nation cannot bet on pure scientists like betting on horses. It can, however, build stables. I remember a Robert Heinlein novel about a research outfit called The Long Range Foundation. The Long Range Foundation was chartered as a nonprofit corporation, dedicated to future generations. Its coat of arms read "Bread Upon the Waters," and it prided itself in funding only scientific projects whose prospective results lay at least two centuries away. It was happy to waste money. Unfortunately, the directors could never do their job right, and the foundation's most preposterous projects quickly began piling up embarrassingly large profits.

The third value I mentioned is the ability to change our worldview. This quality is often subtle, but its importance cannot be overestimated. I think Henry Adams understood the value of pure science when he wrote, in the early 1900s, that Madame Curie's discovery of radioactivity suddenly made the unknowable known. Since ancient times, Western man had worshiped this ultimate material unit called the atom—indestructible, impenetrable, exquisitely unfathomable. Then, at the end of the last century, Madame Curie found that atoms of radium hurled out tiny pieces of themselves, and our view of nature would never be the same again.

It might be helpful to give a couple of examples of this in more detail. I will choose from astronomy, which is the most useless science I know and my personal profession. Actually, astronomy was once highly practical. Early civilizations used it for tracking the seasons, planting crops, and navigation. Since then, astronomy has advanced to its present condition.

As a first example, consider Kepler's discovery that the orbits of the planets are elliptical. Before Kepler, there was universal agreement, dating back many centuries, that the orbits of heavenly bodies were circular. To defer to Aristotle, whose opinions on many things molded the Western worldview, the circle was the natural figure for heavenly motions because of its uniqueness and perfection. Only circular orbits were proper for the divine and eternal planets. In fact, Aristotle arranged the entire cosmos in a sequence of rotating spheres, centered on the Earth. Once nominated, the circle showed great staying power. When people later noticed that the planets changed in brightness—and hence distance from Earth—during their orbits, astronomers invented an elaborate set of circles upon circles, whereby each planet performed a small circular orbit about an imaginary point that itself traveled in a large circular orbit about the Earth. Even Copernicus, who demolished the idea of an Earth-centered cosmos, clung to the idea of circular orbits.

Kepler had the good fortune of being the student of Tycho Brahe, a wealthy Danish astronomer who spent night after night observing the planets from his private island. Brahe's naked-eye reckonings of planetary positions were the most accurate ever taken. Kepler inherited this gold mine of data around 1600. His job was to make sense of it. In addition to having good material to work with, Kepler owed his success to two other factors: he was a dedicated Copernican, and he believed in the Platonist ideal that nature follows mathematically simple laws. What were the laws governing the motions of the planets? What were the shapes of the orbits? Kepler struggled with countless trial orbits of compounded circles. Eventually, he was forced to admit that they just wouldn't fit Brahe's data. Then he discovered ellipses. (Every artist knows the ellipse; it is a foreshortened circle.) One ellipse for each planetary orbit was also much simpler than two circles. The sacred circle had been replaced by the accurate and economical ellipse.

Kepler's success gave strong support to the Copernican system, in which the Earth is simply another planet, orbiting the sun. We know that Newton, as a student, studied Kepler.

When Newton presented his incomparable *Principia* to the Royal Society in London, it was introduced as a mathematical demonstration of the Copernican hypothesis as proposed by Kepler. Newton's *Principia* in turn, with its laws of motion and gravity and its unflagging application of these laws to pendulums and planets, provided a firm scientific foundation for Descartes' view of the universe as a giant mechanical clock. After Kepler and Galileo and Newton, nature became rational.

My second example of how pure science changes our worldview is the fairly recent discovery that the universe is expanding. The galaxies are flying away from each other. When this observed motion is mentally played backward in time, the galaxies crowd closer and closer, stars and planets and even atoms are squeezed together and disrupted, until, some ten billion years ago by the best estimates, the entire contents of the now-visible universe were compressed to a size smaller than an atom. That was the beginning of the universe. It is called the Big Bang.

Virtually every culture in recorded history has had its myths about the origin of the universe and when that origin occurred. Many have believed in no origin at all. Aristotle, for instance, gave numerous philosophical arguments why the universe had to be unchanging and everlasting. One of his arguments went something like this: If the universe had a beginning at some finite time in the past, then there would have been an infinite time before that when the universe did not exist, but had the potential for existing. However, the nonexistent universe could not have slumbered for an infinite time in such a state of pure potentiality. Therefore, the universe has always existed in its present state of perfect composure. Isaac Newton arrived at a similar conclusion by a somewhat more scientific (but still erroneous) approach. Newton argued that if the universe were expanding or contracting, there would have to be a center about which such motion took place. In an infinite space, however, no position in the universe should be so privileged. Therefore, the universe must be always at rest.

The discovery of what the universe is actually doing came in the 1920s. Using a large telescope and various instruments, the astronomer Edwin Hubble was able to determine that other galaxies are moving away from our galaxy, with speeds proportional to their distances. The closer galaxies are retreating from us more slowly than the farther ones. This is exactly the situation for dots painted on the surface of an expanding balloon. From the vantage of each dot, representing a galaxy, it appears that the other dots are moving radially away from it, with speeds proportional to their distances. The view is the same from any dot, and no dot is the center. That was Newton's mistake. Newton didn't realize that expansion could occur about *every* point in space. He didn't have the right picture in his mind. He also didn't have much equipment. I think that if Newton were here now, or Aristotle, or Moses Maimonides, or Francis Bacon, they would sit still for a lecture on the origin and motion of the universe.

It is too early to know the consequences of our discovery

that the universe is expanding. There is no doubt, however, that our worldview has been changed. One sign is that Einstein insisted at first on a static universe—even when his own cosmological equations naturally predicted a universe in motion. For centuries before Hubble, the majestic tranquility of the heavens symbolized the eternal and the immutable. That soothing symbol is now gone.

One can list many other discoveries that are too new to judge. What are the consequences of learning that time flows at a variable rate, depending on the motion of the clock, or that all life forms on Earth get their blueprints from the same four molecules? I don't know, but I am certain that these recent discoveries have begun to seep through our culture and alter our thinking.

Discoveries in pure science are not just about nature. They are about people as well. After Copernicus, we have taken a more humble view of our place in the cosmos. After Darwin, we have recognized new relatives hanging from the family tree. We need to be periodically shaken up. We need periodically to break free from the endless cycle of one generation passing dimly into the next, one human lifetime after another. We got stuck some centuries back, and it was called the Dark Ages. Changing our worldview helps us break free.

I come now to cultural immortality, which, of course, transcends individual nations. To quote Thoreau:

In accumulating property for ourselves or our posterity, in founding a family or a state, or acquiring fame even, we are mortal; but in dealing with truth we are immortal, and need fear no change or accident.

Pure science deals with truth, and there is no greater gift we can pass to our descendants. Truth never goes out of style. Hundreds of years from now, when automobiles bore us, we will still treasure the discoveries of Kepler and Einstein, along with the plays of Shakespeare and the symphonies of Beethoven. The civilization of ancient Greece has vanished, but not the Pythagorean theorem.

Some years ago, I went to Font-de-Gaume, a prehistoric cave in France. The walls inside are adorned with Cro-Magnon paintings done fifteen thousand years ago, graceful drawings of horses and bison and reindeer. One particular painting I remember vividly. Two reindeer face each other, antlers touching. The two figures are perfect, and a single loose flowing line joins them both, blending them into one. The light was dim, and the colors had faded some, but I was spellbound. If our civilization can leave something like that for posterity, it will be worth every penny.

A Modern-Day Yankee

in a Connecticut Court

One day last week, I found in my mailbox the following account from a man I know slightly.

If you look up the Howe family in Hartford, Connecticut, you'll find that a curious item has been passed down among the old family heirlooms. It is a ballpoint pen, found among the personal effects of one Phineas Howe, who practiced law in the last century. The pen is cracked and dirty, but it is unmistakably what it is—a Bic ballpoint. No one living except me knows how Phineas got that pen. Here is my story.

I am an assistant manager of a department store and live in the Boston area. Although I spend most of my time wrestling with inventories, I think of myself as having a decent general knowledge of the world. On the evening of August 9, 1985, I was relaxing at home after a long day at work, when I leaned over to take off my shoes. I must have struck the bookshelf, because my Panasonic home entertainment center came crashing down and hit me in the head.

When I came to, I found myself lying in a meadow, next to a dirt road. Peering down at me was a man in a buggy. He was wearing funny-looking baggy pants and suspenders. As I began to get to my feet, the man spoke to me.

"You from New York?"

"New York?" I repeated, gingerly exploring the bump on my head.

"Yep, New York. I don't know where else clothes like that come from."

"Where am I?" I asked slowly.

The man looked at me as if I were nuts. "You're in back of the Armory," he answered.

"The Armory?"

"The Colt Armory," he said. "In Hartford."

"Hartford," I shouted. "What day is it?" If I'd been out several days, I was in deep trouble with Mr. Godine, my boss.

The man in the buggy shook his head and smiled sympathetically. "It's Monday," he said. "Monday the ninth. Now, why don't you just come along with me to the Armory. We've got a doctor there."

"Have you got a telephone?" I asked quickly.

"Due for one at the first of the year," he said. "Time being, we've got two good telegraph lines."

I silently climbed into the buggy. The man gave a pull on

the reins, his horse gave a snort, and we trotted off down the road.

"By the way," I said, "I know this may sound stupid, but what month is it?"

"August," said my new acquaintance. And then, incidentally, "Eighteen eighty."

Pretty soon, two large smokestacks billowing smoke came into view, then a whole complex of buildings. There were three main buildings, each four stories high, connected together in the form of a capital H. Three sides of the factory were enclosed by a wide dirt road and a wooden fence. The fourth side bordered a river. Through the trees and the buildings, I could just make out the masts of what must have been some large steamboats or schooners at dock.

We rode through the main entrance of the Armory and parked our horse and buggy next to another horse and buggy. Before I had walked ten feet, a crowd of workers, all dressed in baggy pants and suspenders, gathered around and were gawking at me. Either I was crazy or they were, and they had the majority. I made the mistake of being honest. When I told them it was August 9, 1985, the last I remembered, the men roared with laughter. I told them where I worked and where I lived. I began reciting recent presidents: "Nixon, Ford, Carter, Reagan . . . "

"Mind yourself," blurted one burly guy, "or we'll have you carted off to the Retreat for the Insane."

I decided it was time to take a quiet walk around downtown Hartford, so I politely asked for directions and left. By now, I was mostly sure that somehow I had gotten bumped back in time.

It was about a mile and a half to the center of town. On the way, I passed several more horse-and-buggy combinations. I also passed a group of people cheering and hooting as if a contest were about to begin. When I got a better look, I saw that it was a race between a horse and a bicycle. A boy of about twelve, wearing a red cap and striped knickers, straddled the bike and could hardly wait to launch himself. Everyone seemed awed by the bicycle, except two or three older men who were scoffing at it.

I walked on. I have to admit that within a few minutes of wandering around town, I forgot my predicament. It was a warm summer morning and the air smelled sweet. The dirt streets were wide and easy, the traffic was light, and the stores weren't selling the usual. One firm, named Wm. H. Wiley, produced something called over-gaiters. A store called Smith Medicated Prune Company offered free samples. Around the corner were the smokestacks of an enormous Pratt and Whitney factory, advertising machine tools, gun tools, tools for sewing machines, and steam engines, all produced with "precision, durability, and complete adaptation of means to ends." Another company flew a flag with the motto: BETTER MACHINES FOR A BETTER LIFE. Posted against the first-floor window was a drawing of "Thomas A. Edison's new talking machine," showing a long cylinder, mounted at both ends and pressed against some kind of earphone or speaker in the middle. In the picture, a cheerful woman leaned over the machine and was turning the cylinder with a crank.

Getting tired, I spotted a park bench and sat down. I was oddly excited. There was a feeling of progress in the air. Technology was booming. Life was improving.

Then it struck me how to prove who I was. I was a man of the twentieth century. I could reveal to them the wonders of modern technology. They would have to believe me. My knowledge would speak for itself. And there was something else. I'd been taking orders for years. It was about time to be the person in charge. I began feeling a heady sense of power.

The Colt Armory seemed the best place to start, since I already had acquaintances there. I ran all the way and immediately sought out Amos Plimpton, the fellow who had taken me into his buggy. He was operating a metal stamping machine when I found him.

"Mr. Plimpton," I said out of breath, "give me just fifteen minutes with your best machinists. I've got some very interesting things to tell them about the competition. I promise it will be worth their while."

Plimpton miraculously consented, probably giving in to his good Yankee business sense.

After about twenty men had congregated in Plimpton's shop, I got under way. I figured I would start easy, maybe with cars, and work up to VCRs. "Gentlemen," I began, "let me tell you about a very advanced means of transportation called an automobile. I think you've probably got the tools to build one right here in this shop." Silence. I continued. "An automobile has a gasoline engine that revs up when you put your foot on the accelerator, and it will carry you along the road at up to a hundred miles per hour." I smiled.

"How does this gasoline engine work?" asked one fellow.

"Well," I said thoughtfully, "there are cylinders and valves that open and close, and gas and air are brought in and mixed up and ignited by spark plugs."

"*Spark plugs*, uh huh," said another fellow.

The men stood up and began filing out of the room.

"You've got to believe me," I said, flailing my arms.

"What's there to believe," barked one of the workers angrily. "All you've told us are the names of things."

"I'm from 1985. I'm from 1985," I cried out. "I can teach you things."

"Plimpton," somebody said, "call the police. This guy is a loony. The police will know what to do with him."

And that's how it was that I met Phineas Howe. When the police tried to lock me up that first afternoon in Hartford, I demanded a trial. After a terrible scene of kicking and spouting of unknown constitutional amendments, they released me into the custody of Plimpton, who felt some strange responsibility for me. The trial was set for August 16, and Plimpton generously put me up in his house until then. Phineas Howe, without his knowledge and against his better judgment, was appointed my counsel.

A couple of days later, I met Phineas at his office to prepare our case. Phineas was about fifty years old, tall, slightly stoop-shouldered, and paunchy, with a great mop of disheveled hair. His big rubbery face had more skin than was necessary, and he looked sad, like a basset hound. He greeted me at the door with reluctance.

"You the guy from the twentieth century?" He sighed.

I nodded. He let me in. The first thing that caught my eye was the moose head on the wall. I tried to find a clear space to sit down, which wasn't easy. The couch was piled waist high with back issues of *Hunter and Field*, and the cot in the corner was spilling over with shirts and underwear. Papers and food were all over the floor. Finally I located a tiny island of space on the carpet, which belched up a huge cloud of dust when I sat down.

It was blazing hot. Phineas tossed his jacket in a random direction and rolled up his sleeves. "Now," he said, pausing to remove some wax from his ear, "tell me the facts. You realize, of course, what's at stake here. You've been charged with disturbing the peace, attempted fraud, and lunacy."

I repeated my story, while Phineas took down everything on a long yellow pad of paper. I don't think he believed one word I said, but he had been appointed by the public defender's office, and he had his job to do, and he wasn't going to lose this case for any mistake *he* made. As it turned out, Phineas had a shocking track record, but he did have a certain amount of professional pride.

We went through a series of questions and answers, with his asking the questions and my giving the answers. He asked when I was born. "December third, nineteen forty-eight," I replied. "You mean to tell me that you won't be born for sixty-eight years?" he asked in a steady voice. I stopped and figured.

"Yes, that's right," I said. "I understand. I understand," Phineas said with a pained expression, and scribbled on his yellow pad. It went on like that for half an hour.

The seriousness of my situation began sinking in. "I wouldn't have been in this mess if the machinists at Colt had given me ten more minutes," I said glumly.

"Those people weren't set up for someone like you," said Phineas with an impatient brush of his hand. "I was thinking of trying to get Tom Edison down here as an expert witness. I was part of a law firm that helped him in a patent suit a few years ago. You know about Edison, don't you?"

I nodded appreciatively.

"Edison is bright enough to figure this thing out," Phineas said, and then added, "one way or the other."

After some hasty inquiries, we walked to the telegraph office and wired Menlo Park, New Jersey, where Edison worked night and day in his lab. An hour later we got a return message, which Phineas wouldn't let me read but which said something to the effect that we could go to hell and back unless I could help Edison with his lighting system for New York City. My attorney looked at me searchingly, and I said, "Certainly." This was no time to lose confidence. Phineas wired back that single word, and, within another hour, we received a second message saying that Edison would be arriving on the New York & New England line at 10:13 A.M., August 16.

The trial was held in the Court of Common Pleas, in the new brick County Building on the corner of Trumbull and

Allyn. Plimpton had wanted badly to be there with me that morning, but his daughter had suddenly become very sick with pneumonia. As I was leaving his house, I mentioned the possibility of making some penicillin, but I didn't know more than the name of it. Plimpton stared at me blankly and I left. I had grown fond of him and his wife and felt terrible about their daughter.

Phineas arrived at court looking as if he hadn't changed clothes for forty-eight hours. He was carrying several yellow pads and an armload of *Popular Science Monthly*s.

"Don't say anything except when you're up on the stand," he whispered urgently to me, "and then answer only direct questions. I'm on top of this thing." I nodded and followed him to our seats. "We're not giving those buggers one inch," he whispered again, "especially that starched son-of-a-bitch Calhoun."

"Who is Calhoun?" I whispered back. For an answer, Phineas simply glared across the room at a calm man in his late thirties, then opening a trim leather briefcase. That was Thomas Calhoun, the prosecuting attorney. He was flanked by two young assistants. All three wore immaculate gray suits. Calhoun was slender and had very black hair. Calhoun was the type of person who never utters a word that isn't right. He got his law degree from Yale. I saw all of this in the first five minutes and became depressed.

Then Judge Renshaw walked in and everyone stood up. "Renshaw doesn't have the brains to bait a fish hook with," whispered Phineas. The trial began.

Besides the parties involved, about twenty spectators had come over from the Aetna Insurance Company and sat in the back of the courtroom. Throughout the trial, they were continuously fanning themselves with paper fans advertising the Spring Grove Funeral Parlor.

I won't repeat the opening remarks. Calhoun presented the case for the town of Hartford, bringing in several men from the Armory to testify. He was brief and smooth as silk. Phineas stated our position. Judge Renshaw, a small, quiet man, seemed puzzled by the whole thing and said nothing.

Then Edison arrived. "Where's Phineas Howe?" he boomed, walking down the center aisle. The bailiff started to intercept him, but the judge raised his hand. There was a reverent hush, as everyone turned to get a glimpse of Thomas Alva Edison. Finally, a court attendant led him to our seats. Edison was a barrel-chested man with burning blue eyes. He gave me an odd look and said to Phineas, "Make this goddamned fast. I'm leaving on the twelve thirty-three."

Phineas promptly had me put on the stand and introduced Edison as his expert witness. "I intend to prove," announced Phineas, "that my client has knowledge of a technology so far advanced beyond ours that he could only be a citizen of the late twentieth century. Or beyond. This knowledge will be confirmed by the leading inventor of our age, Mr. Thomas A. Edison." The people from Aetna momentarily laid down their fans and clapped. Calhoun, to my satisfaction, shifted uncomfortably in his chair and began whispering to his aides. Phineas and I now held the trumps.

"To begin with," said Phineas, turning to me, "tell the court about your house."

"Well," I said, "I have a refrigerator, a dishwasher, a stereo, a tape deck, two telephones, a television, a video cassette recorder, a microwave oven, a personal computer, and a Chrysler in the garage." Actually, I was a little embarrassed about flaunting my prosperity in front of the court like this, but Phineas had insisted. Then Phineas got me to explain what each of these items did.

"Objection," said Calhoun. "The defendant has merely invented a lot of fancy names and functions. He is wasting the court's time."

"I believe Mr. Edison will determine that," said Judge Renshaw. "Objection overruled." The judge looked expectantly at Phineas, who was rapidly paging through one of the *Popular Science Monthly*s.

Then Phineas asked me to explain to the court and to Mr. Edison how a television works.

"A radio signal comes in from a broadcast station," I began, "and is picked up by the television antenna. This signal then goes into the television and directs electricity at a picture tube, which has lots of dots on it. The dots light up when the electricity hits them. That's what makes the picture."

Edison was champing at the bit.

"Your honor," said Phineas, "will you allow Mr. Edison to question the defendant?"

Judge Renshaw nodded.

"Are there wires that go directly to this picture tube?" asked Edison.

I thought hard. "I don't think so," I answered.

"What did you say?" asked Edison.

"I don't think so," I repeated.

Edison looked as if he still hadn't heard me.

"I don't think so," I shouted.

He nodded, and said, "In that case, don't this picture tube need a vacuum inside?"

"Sounds reasonable to me." I looked over at Phineas. His hands were over his eyes.

"Does the television use direct or alternating current?" asked Edison.

I thought again. "Well, I think it comes out of the wall alternating." There was a round of laughter in the courtroom. Phineas still had his hands over his eyes, but seemed to be peeking through his fingers.

"What's that you said?" asked Edison. He was definitely hard of hearing.

"I think it comes out of the wall alternating," I shouted.

"But is there a transformer or rectifier?" asked Edison.

"What's that?" I asked.

"A transformer increases the voltage and decreases the current, or vice versa, keeping the product constant. You lose more power with low voltage. A rectifier changes alternating current to direct current. I've been having a hell of a time with my transformers for Pearl Street. The capacitances aren't right."

What Edison was saying was extremely interesting. "Now tell me about this picture tube," he continued. "You say it lights up when electricity hits it?"

I nodded.

"How does that happen?" asked Edison. "What's this picture tube made of?"

I moved on to refrigerators. "Now a refrigerator is a marvelous machine," I said. "It keeps food cold with electricity. You can forget about hauling around big chunks of ice."

"How does one of these *refrigerators* work?" Edison asked.

"There's a motor inside," I answered loudly. "It shoves the heat outside of the refrigerator." To my surprise and embarrassment, I discovered I couldn't explain much more about refrigerators, although I was sure there wasn't much to explain.

Edison looked at his watch.

"TNT," I said firmly. "It's a very powerful explosive and excellent for weapons." The people from Aetna stopped fanning. "Tri-nitro something," I added.

"You mean nitroglycerin?" asked Edison.

"No, not nitroglycerin. Tri-nitro something."

"What are the ingredients?" asked Edison.

"Nitrogen is one," I answered.

Edison looked at me contemptuously and said, "I don't think this fellow knows one goddamned thing about technology of any century. And he's certainly no help to me." At that, he strode out of the courtroom. Phineas, clearly shaken, told me I could sit down. Calhoun looked smug. I felt humiliated.

Judge Renshaw cleared his throat and indicated that it was time for the closing remarks. Calhoun went first.

"I believe it is clear," he said evenly, "that the defendant has demonstrated no knowledge of advanced technology, no proof that he is in fact a citizen of the twentieth century. I ask the court, therefore, to proceed on the basis that he is either a deliberate fraud, and has tried to deceive the honest people of our town, or a dangerous lunatic. The prosecution recommends five years' incarceration in Lockwood Prison, or an equal period in the Connecticut Retreat for the Insane, whichever is appropriate."

Then it was our turn. In a daring move, Phineas asked to have me put in the witness box one last time. He walked over to me, smiled, and said quietly, "Do your friends in the twentieth century know how televisions and automobiles and computers work?"

I was keenly aware of being under oath. "There's a fellow from Acme Electronics in Cambridge who fixes my television when it's busted," I said, "but I couldn't say I really know him." I thought. "When I lived in Watertown, I knew someone who could build an automobile brake system from spare parts. Computers, well . . . " I shook my head no. "It's recommended not to take them apart."

"Do you mean to tell me," said Phineas, almost whispering, "that only a handful of people from your century understand how these things work?" Phineas was mocking me, probably to inflate his own lifeboat as the ship was sinking. I didn't blame him much, especially since he was right, but it made me mad.

"Do *you* know how a *telegraph* works?" I asked Phineas, angrily.

"I object," said Calhoun, leaping to his feet. "The knowledge and credibility of my colleague, Mr. Howe, are not relevant here. Also, it is highly improper for a defendant to argue with his own counsel."

"Objection sustained," said Judge Renshaw, yawning.

"Do *you* know how a telegraph works?" I said to the judge.

Phineas led me quickly back to our seats.

"Does the public defender have anything more to say?" said Judge Renshaw.

Phineas was writing rapidly, using a chewed-up pencil he'd whittled down to nothing with his pocket knife. "You got anything to write with?" he whispered frantically.

"Sure," I answered, and took a ballpoint pen out of my pocket. He grabbed it and continued scribbling without looking up. Suddenly he stopped and stared at the pen. He pushed the button and watched the point go in. He pushed it again and the point came back out.

"Son of a bitch," he said softly. "Will you look at this?"

He got up, holding the pen, and walked over to the judge's bench. After some animated mumbling, the judge motioned for Calhoun to come over. Then the judge asked to see all of us in his chambers.

I was acquitted, of course, although the people from Aetna never knew quite why. There was a minor sensation following the trial, and a reporter from the *Hartford Times* came out. He brought a photographer with him. It seems he wanted to get a picture of me riding a horse. "The man from the twentieth century, temporarily inconvenienced, gets about by horse," or something like that.

A small crowd of people had gathered across the street from the County Building, and the reporter was there and the photographer and Phineas, and I mounted up. I guess I mounted that horse on the wrong side, because the next thing I knew I'd been pitched and was sailing by L. T. Frisie and Sons, head-first toward a lamppost. That was the last I saw of old Hartford.

When I awoke, I was lying on the floor of my living room, covered with dust. My wife was bathing my forehead with a wet cloth. She sighed with relief as I opened my eyes. "Dear, where have you been?" she said. "I couldn't find you for over an hour, and then I heard a loud thump and found you like this, unconscious." Despite my headache, I managed a smile.

The Origin of the Universe

For the last couple of decades, physicists have been pushing their theories of matter and energy backward in time, closer and closer to the primal explosion that started the universe, some ten billion years ago. Indeed, it is common these days for scientists to hold forth on "The Origin of the Universe." When you go to such lectures, however, you soon learn that you're not getting The Origin, but a billionth or a trillionth of a second later.

And so it was that I casually took my seat at yet another lecture titled "The Origin of the Universe," given at Harvard in the spring of 1984. The lecturer was the British scientist Stephen Hawking. The hall was packed. Hawking, then forty-two, has become one of the seminal theoretical physicists of our time. He has also suffered for years from a worsening motor neuron disease, which has ravaged his body but spared his mind. On this afternoon, as Hawking sat in his wheelchair, laboring to utter a series of sounds that were translated into words by a student, I gradually realized what I was hearing: Hawking had traveled back the whole distance. For the first time, a preeminent scientist was tackling the *initial* condition of the universe—not a split second after the Big Bang, as I'd heard about before, but the very beginning, the instant of creation, the pristine pattern of matter and energy that would later form atoms and galaxies and planets.

In any other circumstance, physicists would hardly think twice about discussing initial conditions. The "initial conditions" along with the "laws" are the two essential parts of every model of nature. The initial conditions tell how the particles and forces of nature are arranged at the beginning of an experiment. The laws tell what happens next. Any predictions rest on both parts. Set a pendulum swinging, for example, and its motion will be determined by the initial height where your hand let go, as well as by the laws of gravity and mechanics. But Hawking's pendulum is the entire universe. And he is attempting to reason out what theologians and scientists alike have previously assumed as a given. He is attempting to *calculate* where the hand let go. Hawking's equations for the initial state of the universe, together with the laws of nature, could predict the complete outcome of the universe. They could tell us whether our universe will expand forever or reach a maximum size and then collapse. They might explain the existence of planets, or of time.

How could anyone know whether Hawking's equations are right? Could the human mind even grasp Creation? And, just as puzzling, how did science arrive at such outrageous self-confidence? I asked these questions to myself and a dazed colleague as we wandered out of the lecture hall and away from the campus, walking past cars and children in mittens.

Physicists today are not modest—and with some reason. Just in this century, they have discovered and successfully tested a new law for gravity, a theory for the strong nuclear force, and a unified theory of the electromagnetic and weak nuclear forces. They have proposed further laws that might unify all of the forces of nature. Physicists have demonstrated that time doesn't flow at a uniform rate and that subatomic particles seem to occupy several places at once. These victories, often in territory far removed from human sensory perception, have created a strong sense of confidence.

In many of the more heady advances, theory has outdistanced observation, let alone application. For example, the unified theory of the electromagnetic and weak nuclear forces, developed in the 1960s, predicted the existence of new particles that weren't discovered in the laboratory until the 1980s. Superdense stars, fifteen miles in diameter, were predicted in the 1930s—more than thirty years before they were first observed in space, thousands of light-years from Earth. Einstein's general theory of relativity predicted that a ray of starlight passing near the sun should be deflected five ten-thousandths of a degree by the sun's gravity. When a novel experiment confirmed this minuscule effect several years later and Einstein seemed blasé, a student asked him what he would have done if his prediction had been refuted. He answered that then he would have been sorry for the dear Lord, because "the theory *is* correct."

With such confidence, physicists have grown accustomed to extrapolating their theories to situations that could not possibly be witnessed by human beings. Hawking's work on the beginning of the universe is an extreme example of this kind. In the evolution of the universe as a whole, gravity is the dominant force to be reckoned with. Hawking has extrapolated Einstein's theory of gravity back to an epoch that was not simply prior to life, but prior to atoms. Stranger still, the early universe was of such high density that its entire contents, including the geometry of space itself, behaved in the hazy, hard-to-pin-down manner of subatomic particles. The methodology needed to describe such behavior is called quantum mechanics, and the application of this methodology to gravity is called quantum gravity. According to the theory of quantum gravity, it was possible for the entire universe to appear out of nothing.

Hawking has mathematically investigated the kind of universe that could have appeared out of nothing. Would the infant universe be finite or infinite in extent? Would it curve in on itself? Would it look the same in all directions? Would it be expanding rapidly or slowly? The answers to these questions are buried in a difficult equation. That equation will likely take a long time to solve, an even longer time to test against its predictions, and it may be plain wrong. Nevertheless, it reflects an extreme confidence in the power of human reason to reveal the natural world. Hawking, like Darwin, has ventured into regions previously forbidden to human beings, and to scientists in particular. Hawking's work, right or wrong, is a celebration of human power and entitlement to knowledge.

Men and women have always longed to understand and control their world, but they have constantly met with obstacles. In different ages and cultures, they have tried different means to clear their path—magic in primitive cultures, religion and science in more evolved societies. Primitive man believes in his power to control nature and other men through magic. He believes he can make rain by climbing up a fir tree and drumming a bowl to imitate thunder. He believes he can bring a cool breeze by wrapping a horsehair around a stick and waving it in the air. But with growing experience, man realizes that these methods have their limitations. Rain and cool breezes don't always come when requested. At this stage of development, as the anthropologist Sir James Frazer says in *The Golden Bough*, man stops relying on himself and throws himself at the mercy of higher beings. Thus begins religion. And a surrender of personal power. My rabbi once told me that man has always made of God what he wished to be himself.

But, as man increases his knowledge, this new reckoning with the universe also needs revision. For the gods, reflecting the ignorance and superstition of man, have often been given human personality along with their powers. The gods get drunk, as the Babylonian deities did on the night before Marduk went out to do battle against Chaos. They are jealous and spiteful, like Hera, who destroyed the Trojan race because she had placed runner-up in a beauty contest judged by a Trojan. If natural phenomena are controlled by such gods, then those phenomena should be subject to whim and to passion. Yet the more man studies nature, the more he finds evidence for regular laws. Seasons repeat, stars move on course, and stones fall predictably. The study of these regularities marks the method of science. Through science, man regains much of his primitive confidence in self-power, with control now replaced by knowledge. Knowledge is power. Man might not be able to control the weather, but he can try to predict it.

With the beginning of modern science in Europe, people measured eclipses, they dissected cadavers, they observed mountains on the moon with the new telescopes, they peered at lake water through microscopes, they studied magnets and electricity. Copernicus declared that the earth went around the sun. Paracelsus announced that disease was caused by agents outside of the body, not by internal humors. Galileo pointed out that moving bodies maintained their motion unless acted on by external forces.

Yet flowing deep through human culture remained the idea that some areas of understanding were off limits or beyond mortal grasp. Adam and Eve were punished because they ate from the forbidden tree of knowledge, which opened their eyes and made them "as gods." In *Paradise Lost*, Adam asks the angel Raphael to explain Creation. Raphael reveals a little, and then says:

... the rest
From Man or Angel the great Architect
Did wisely to conceal, and not divulge
His secrets to be scann'd by them who ought
Rather admire. ...

Doctor Faustus approached other authorities for knowledge and had to pay with his soul. People also questioned to what extent the universe was subject to human rationality. Descartes had likened the world to a giant machine, but many viewed such reductions as threats to the power of God. In his Condemnation of 1277, the bishop of Paris made it clear that no amount of human logic could hinder God's freedom to do what He wanted. Even Isaac Newton, master logician and reductionist, surveyor of all natural phenomena, comes to the end of the *Principia*, the General Scholium, where he lets down his hair and confesses that the synchronized performance of moons and planets could never be explained by "mere mechanical causes," but requires "the counsel and dominion of an intelligent and powerful Being." Furthermore, it would be impossible for mortal man to fathom the art of that divine balancing act: "As a blind man has no idea of colors, so we have no idea of the manner by which the all-wise God perceives and understands all things." Newton, both scientist and believer, was caught between his own power of calculation and the unknowable power of God.

But the unknowable continued to beckon, and man, although fearful of lifting the veils, was still driven to try.

After Newton, a great debate arose over whether the solar system could be explained on a rational basis. Similar debates echoed through later centuries. In the eighteenth and nineteenth centuries, geologists argued over whether changes in the Earth came about through gradual transformations, obeying natural law, or through sudden catastrophes, ordered by a tampering God. The mode of thinking in the late 1800s, just before Madame Curie found that the sacred atom could be splintered, was described by Henry Adams in this way: " . . . since Bacon and Newton, English thought had gone on impatiently protesting that no one must try to know the unknowable at the same time that every one went on thinking about it."

For Henry Adams, the unknowable that became known was the atom. For modern biologists, it is the structure of DNA and possibly the creation of life. For modern astronomers, it is the distance to the galaxies and the shape of the cosmos; for modern physicists, perhaps the grand unified force and the birth of the universe. Layer by layer, the unknowable has been peeled away, and examined, and made rational. Scientists today, humbled by their dwindling size in the cosmos but emboldened by their success at adjusting, have staked out all of the physical universe as their rightful territory. And they intend to let their theories and equations take them to places they cannot go with their bodies. In the introduction to one of his recent papers, Hawking says: "Many people would claim that the [initial] conditions [of the universe] are not part of physics but belong to metaphysics or religion.

They would claim that nature had complete freedom to start the universe off any way it wanted. . . . Yet all the evidence is that [the universe] evolves in a regular way according to certain laws. It would therefore seem reasonable to suppose that there are also laws governing the [initial] conditions."

To me, Hawking's work, although strikingly bold, is a natural extension of what science has been doing for the last five hundred years. But the question remains: After physics has reduced the birth of the universe to an equation, is there room left for God? I posed this to a colleague who has done significant calculations on the origin of the universe and is also a devout believer in God. He answered that while physics can describe what is created, Creation itself lies outside physics. But with your equations, I said, you're not giving God any freedom. And he answered, "But that's His choice."

How the Camel Got His Hump

It is early evening and I am putting my daughter to bed. She sits beside me in her yellow pajamas, with her head against my shoulder. For the third time, we are making our way through the *Just So Stories*. My daughter wants to know if the magic Djinn in charge of All Deserts could really cause the Camel's back to puff up so suddenly, and what good is the hump anyway. She has asked this before. Tonight, I am prepared, having looked up camels in the library and talked to some knowledgeable friends. The hump, I explain, is made of fat, which all animals need to live on when they can't find food. The camel keeps all its fat in one place, on its back, so that the rest of its body can cool off more easily. Staying cool is important in the desert. The penguin, on the other hand, needs to stay warm, and spreads its fat in a thick layer all over its body, like a blanket. I tuck my daughter in.

"Daddy, camels are wicked smart, aren't they?" she says, yawning.

"Not really," I say. "Camels didn't figure things out on their own. Nature spent millions and millions of years working on camels, making lots of mistakes until they came out right."

I turn off the lights. The streetlamp outside shines through the bedroom window. I think of my visit to New York last week, coming into the city at night on a bus, with the buildings and towers all lit up, slender and beautiful and fragile, like miniatures. And then, on the Queensboro Bridge, with the streetlamps passing one by one, the light pulses on the vinyl seat in front of me, making it look like throbbing skin, the very thin skin on a person's throat, quivering with each pulse of blood in the veins underneath.

My daughter sneezes. "Guess what I made in school today, Daddy." she says.

"What?"

"A Pilgrim, for Thanksgiving. And before that, I climbed up to the fourth rung of the ladder. The fourth rung. Mrs. Gauthier saw me."

I kiss her and walk to the window. "Come look at the moon with me," I whisper. She gets out of bed and tiptoes barefoot across the carpet. We open the white shutters.

"Men have gone to the moon and walked on it," I say. "Just a few years ago."

The night is broken by the sound of a car down the street.

I look at the moon again, hanging in space, and I imagine giant wheels of steel, rotating silently in the darkness over-head. I imagine thousands of satellites whizzing around the planet in all directions, narrowly missing each other. I imagine smooth cylinders suddenly launched upward, lighting the night with the red fire from their engines, arcing toward cities. New toys of new creatures. And below, the ancient Earth waits.

"Back to bed," I whisper to my daughter. I tuck her in again, folding the blanket carefully across her chest.

"Daddy," she says, "will you read to me again about the Djinn, and how he made the hump puff up with magic?"

"Another night," I answer.

IRONLAND

One evening not long ago, as I sat quietly reading by the fireside, a hooded stranger knocked on my door, handed me a crumpled missive, and quickly left. I would not have placed much credibility in his curious story, which I reprint below, had I not later noticed the wood was crushed where he brushed the door frame.

I have been walking the streets for days, virtually blind, in search of another creature of my kind. How I came here through the vastness of space I cannot tell, but I feel compelled to share something of my home. I will call our world Ironland, not because we call it so, but to make its nature clearer to you. In Ironland everything is made of iron. No other elements exist. Picture a land with no air, no rain, no grass; no oxygen or hydrogen or carbon. Imagine a planet with only iron and what can be fashioned from iron. Knowing no other way of life, we consider this state of affairs perfectly natural.

Our world is far simpler than yours. First of all, chemistry is unheard of, there being no other elements available to react with iron. I have been astonished at the myriad chemical phenomena you have: photosynthesis, battery power, taste. Not even our science-fiction writers have imagined such things. Still, we enjoy some compensations. You will grasp at once how durable our structures are, compared to yours. Without corrosion and decay, our splendid houses stand forever and maintain their initial gray-white color. I would think that architects in your world, especially those anxious that their buildings last a thousand years, must find oxidation a frightful nuisance. Indeed, aging of all kinds proceeds more swiftly on a world with chemistry, although this still does not explain why your creatures rarely live beyond one hundred Earthly years.

You may wonder what distinguishes the animate from the inanimate in Ironland, and I will tell you. We experience nature on its simplest terms and have decided that life, in its essence, consists of information—and mechanisms for expressing that information. Now iron, as you know, has magnetic properties. Some years ago, our scientists discovered that the microscopic magnetic regions inside our living matter are oriented according to definite patterns. That is, if you think of each of these regions as being a magnet, then the little north poles in animate matter point up or down in very particular arrangements, analogous to the sequences of dots and dashes in your Morse code or the on and off switches in your computers. Any piece of information can be reduced to such a sequence and stored. In lifeless forms, like rocks and hammers, the tiny internal magnets point haphazardly, with little relationship to each other. No greater magnetic information resides within a rock than in a word of letters taken randomly from the alphabet.

Magnetism in our society is akin to money in yours. We base our status on it. But the Board of Magnetometers, curse them, has more or less dictated the system. The lower classes, like welders, are permitted an overall magnetic field of no more than 100 gauss. (So that you will understand me, I have converted our magnetic units to yours. One gauss, if I am not mistaken, is about twice the magnetic field strength of your planet.) The middle classes, like sculptors and doctors, are allowed up to 1,000 gauss. Some members of the upper classes—I know one politician in particular—boast magnetic fields as high as 10,000 gauss and more. Now, I'll take you into my confidence, but this must never get back to my world. Some of us have noticed that with increasing status comes increasing stupidity. Eventually we realized why. You see, to get high overall magnetization, the microscopic magnetic regions within a person must line up, with most of the little north poles pointing in the same direction. Otherwise, they'll partly cancel each other, reducing the overall magnetic strength. But as a greater number of the microscopic magnets are restricted in their orientation, less are available to store information. It's like restricting a larger and larger fraction of the letters in a word to be the single letter *a*. The extreme case is when all the microscopic magnets point in the same direction, producing a maximum magnetic field of about 20,000 gauss. At this point, all intelligence has been abandoned in favor of status.

I myself carry around 300 gauss, which in my opinion is enough to make ends meet but not so much as to go to my head. I am a writer. Often, I have felt grateful for this one modest talent, as I am somewhat homely and definitely lacking in social graces. My dear mate was recently certified at 310 gauss, although she deserves more. I noticed her fine mettle the first day we met, at the foundry. When I say noticed, you must appreciate that all our sense perceptions are magnetic and operate on much the same principles as some of your metal detectors.

I suppose I'd better explain something about our sexuality. Roughly speaking, your maleness and femaleness correspond in our land to the north and south poles of a magnet. But since any magnet has both poles, every person in Ironland is bisexual. Depending on how you're standing or sitting in relation to someone else, you can find that individual extremely attractive or repulsive. As you can imagine, courtships have to be handled with great delicacy, and you can still slip into an awkward position after many years of marriage.

There is a saying in our world that wayward spouses can usually be turned around. But sometimes one encounters a whole group of disagreeable, misdirected people, and that leads to war. Regrettably, warfare in Ironland suffers for want of strong weapons. Without chemical reactions, we lack chemical explosives. Oh, what I could do with some of your gunpowder or TNT back home.

Much worse, we've failed miserably to build nuclear explosives. I must admit, however, the reason is not without its fascination. As you well know, the particles in atomic nuclei interact with two kinds of forces: a repulsive electrical force, acting between the protons, and an attractive nuclear force, acting between both protons and neutrons. The first force is like a compressed spring while the second like a stretched spring, each poised to snap back to its natural position, releasing energy in the process. Unfortunately, the two kinds of springs pull in opposite directions, so when energy is gained from one it is lost from the other. To get an explosion, of course, more energy must be released than absorbed. Your fission bombs produce energy by splitting nuclei. This method only works for the heavy nuclei, like uranium. On the other hand, for light nuclei, such as hydrogen, net energy is produced by joining them together. You call weapons made in this way fusion bombs. Now that I understand these things, it is not so remarkable there should be a special atomic nucleus unluckily caught right in the middle—neither light enough to yield net energy by fusion, nor heavy enough to do so by fission. In effect, its stretched and compressed springs completely cancel themselves in either direction. That barren and singular nucleus is none other than iron, the sole element of our world.

Nevertheless, being people of some intelligence and

resourcefulness, we've found ways of doing away with each other. One can always heat an enemy to oblivion. When the temperature of iron exceeds 768 Celsius, still well below its melting point, the substance loses all its magnetism. Under such heat, the tiny internal magnets become totally disoriented. It is death by loss of all knowledge, sense of self, and status—but without destruction of material. Rather humane, don't you agree?

We are a cultured people. Our poets cannot write of oceans, but they have mused on the latent stillness of low temperatures, the texture of a cubic lattice shifted, the inner seething of magnetic storms. Our artists cannot paint, but they have created winding sculptures whose forces tingle helically. Confined to the most primitive form of the material universe, we have yet risen to great heights of expression.

Now you know a bit about Ironland. I would write far more, but time does not permit. Already, I am rusting in your wretched air and must depart. Good-bye.

OTHER ROOMS

DURING THE YEARS I was preparing for a career in science, preoccupied with the autonomy of equations and instruments, human beings continually inserted their idiosyncrasies into my education. John, a boyhood friend and partner in early scientific adventures, first exposed me to the notion that success may arrive by a side-door approach. His gadgets worked and mine didn't. He never saved the directions that came with new parts, he never drew up schematic diagrams, and his wiring would wander drunkenly over the circuit boards; but he had the magic touch, and when he sat down, cross-legged, on the floor of his room and began fiddling, the transistors hummed. Watching over his shoulder in an attempt to discover why things worked in his house and not in mine was totally useless, nor was he able to explain anything. John didn't squander his time learning theory.

Afternoons, I would drop over from school with some idea I'd read in *Popular Science* or an interesting scheme of my own and often find John stretched out and spindly on his bed, brooding over his most recent batch of poor grades. At the mention of a new project, with the barest shred of description from me, he would perk up and boost off, like a jazz musician transforming a ratty tune into a pulsing, living cascade of sound, never writing down a single chord. He'd instantly begin pulling electrical wires, soldering irons, chemicals, or whatever he thought was required from boxes piled here and there (in retrospect, my own room suffered from too much neatness) and, with the usual Bob Dylan record howling in the background, get down to business. Soon we'd be adrift, abandoning books, formulae, and schoolwork in favor of wonderful scientific devices, at great distance from the tiny voice of John's mother calling him to dinner.

Our most successful collaboration was a project we entered in the county science fair during my junior year of high school. It was a communication device, but the novel feature was the absence of electrical wires connecting transmitter to receiver and the use of light alone for encoding and transmitting sound. When a person spoke into the transmitter, the sound vibrated a stretched balloon, upon which was mounted a bit of silvered glass. Light originating in the transmitter and reflected from the little mirror thus carried along, through its variations in intensity, information about the original sound vibrations. This information was reconverted into sound by a photocell and amplifier at the receiver. For several months we had ravaged hardware and electrical supply stores around town and were intensely pleased with our final product. The day before the judging, after numerous and brilliant trials at up to fifty feet separation between talker and listener, something broke. I went home in a state of depression. Two days later I received an astonishing call from John saying that he had carried our stricken contraption to the fair late that night and craftily connected the transmitter and receiver with a wire run under the table. Next day the judges were fooled, apparently, and awarded us first prize. John was all practicality.

When I lived in Watertown, more than a decade later, a piano tuner named Phil used to come about every four months to spruce up our upright. Phil, I gradually learned, was just as theoretical as I but free of the nuisance of dressing his dizzy flights with scientific respectability. He was full of ideas about the origins of the universe, what might have come before, and other cosmological subjects. You could never tell, as he spun his fantastic monologues dancing back and forth between science and philosophy, which of his ideas he had read somewhere and which he was creating out of his head at that moment.

After his second or third visit, Phil somehow got wind of the fact that I was a physicist, cementing our relationship and thereafter increasing the tuning job from one to two hours. Upon his entrance into my apartment, Phil would briefly sound out the keys for about fifteen minutes, as if on a dry run, and then disengage his bulky frame from the piano bench and begin spouting his latest theories. My favorite was one in which the solar system was really a big atom, at least as perceived by some giant, unspecified beings, and further out in space the galaxies formed a galactic solar system, which was really a galactic atom, and further out a number of different universes orbited each other. As Phil would describe this hierarchy of orbiting worlds, his voice increased appropriately, his gestures expanded into large arcs that narrowly missed the chandelier, and both of us would be irresistibly lifted up and away into outer space, my Watertown apartment and the entire neighborhood becoming a vanishing speck in the cosmos. Such fantasies would sometimes go on for half an hour. Phil had stamina. In the interlude where he caught his breath, I would sometimes come to my senses and attempt to inject some science into the discussion, drawing on my professional resources. After a fleeting look of worry, Phil would brush off this unwelcome bit of goatishness on my part and continue to greater heights.

On occasion, Phil had the facts exactly right, as far as I could tell, and my approving nod would bring forth a bright grin. One such time was his sobering description of how the continual creation and destruction of microscopic black holes distorts and punctures space, leading to a foamy structure at the most minute levels of reality. That day it took us nearly three hours to get the piano tuned, and I was late to the university.

During Phil's last visit, which was shortly before I moved, I had to immediately excuse myself to the study after letting him in, for I was behind in my research and had to make a presentation later in the day. Phil took this rather grumpily and, for a time, pounded the keys with unnecessary vigor. Without the intellectual excursions, in fact, he accomplished his work in a scant sixty minutes. As I sat working at my desk in the next room, Phil burst in without warning and, viewing the clutter of equations and books, said, "I see you do things the long way."

About a year later I learned with sadness that Jon Mathews, forty-eight, a former professor of mine in graduate school, and his wife had been reported lost at sea while crossing the Indian Ocean alone in a thirty-four-foot sailboat. Such an unkempt catastrophe simply could not be fastened onto other memories of Mathews, who was as meticulous in his scrubbed, crew-cut appearance as he was in his tidy mathematical calculations and who had seemed, at least to us students, to always look twice before leaping. Perhaps it was exactly his cautious and careful style that denied Mathews any truly outstanding contributions to his subject, theoretical physics. Something was missing, some touch of irreverence or deepness or momentary lapse of the rules. In a paper he is remembered for, titled "Gravitational Radiation from Point Masses in a Keplerian Orbit," Mathews begins, "One might expect masses in arbitrary motion to radiate gravitational energy. The question has been raised, however, whether the energy so calculated has

any physical meaning. We shall not concern ourselves with
this question here. . . .", and then goes on to do a textbook
calculation of the conjectured effect. In fact, Jon was a superb teacher and a coauthor of the widely used textbook
Mathematical Methods of Physics. At the blackboard he was in
his element, distilling the physical world into beautifully
chalked equations running on for many feet, explaining each
concept in mechanics or electromagnetism so clearly and
exactly that you could begin to see an equation on the board
wiggle back and forth like the pendulums or springs it
described. Jon was such a good teacher that some of us au-
dited his courses without credit, forcing a space in our fantastic graduate studies to hear a familiar subject discussed with
elegance and precision. I became friends with Jon through our
common love of sailing. Even then, years before his ill-fated
voyage, he dreamed of one day sailing around the world,
talked about it, and recruited eager students to serve as crew
for weekend jaunts on his boat.

A cruise I remember was to Catalina Island, California,
about thirty miles off the coast of Long Beach. Jon's wife, Jean,
wasn't along, but one of his children was. Going out, we had a
stiff and shifty breeze. Jon hustled around the boat with surprising agility, cranking winches, reefing the main, resetting
pulleys, and generally in complete control, as calm as if he
were solving a straightforward boundary value problem at the
blackboard. His boat was immaculately kept, its brass hardware gleaming and in good repair, lines religiously coiled
when not in use, and every item in its proper place—a rare

state of affairs on sailboats. That night, at anchor off the island, with the boat gently rocking and us cozy in the cave-like cabin within, Jon unwrapped his most recent toy to show me, a sextant, his voice pitched high as always when he was explaining something new. We never talked physics on board. I don't know why, we just never did.

The physicist Freeman Dyson has likened much of scientific research to craftsmanship, in which "many of us [scientists] are happy to spend our lives in collaborative efforts, where to be reliable is more important than to be original." That kind of science, even done well, was not enough for Jon. He pursued one hobby after another to complete himself, including an unexpected mastery of Eastern languages, revealed one day in his orderly office when he pulled down from the bookshelf a volume in Sanskrit and began reading to me.

This was all early 1970s. Little by little, Mathews garnered the necessary expertise, the many practice cruises, and the navigational and geographical knowledge to materialize his dream of sailing around the world. On his 1979–1980 sabbatical leave he set sail from Los Angeles in early summer, heading west. The last reported radio contact with him was in December 1979, several hundred miles from Mauritius. A friend was waiting in South Africa for his arrival, which never occurred. Mathews and his wife were apparently the victims of an Indian Ocean cyclone, the counterpart of hurricanes in the Caribbean or typhoons in the China Sea. The peak season for such storms in the Indian Ocean begins in December, and

Mathews had planned his trip so that the crossing would take place well before. However, according to reports I've read, he was already behind schedule before reaching Australia and, rather than turn back or remain berthed for an additional six months while the storm season passed, decided to risk the crossing.

I have tried to re-create in my own mind what must have been happening in his, as Jon pondered charts and timetables and weather reports in Australia and then decided to take the biggest gamble of his life. I picture him as the storm first hit, suddenly finding himself out of control in a vast, raging place he'd never imagined, a room without walls or ceiling or solution. Jon's brilliant lectures in the classroom, his beautifully scripted Sanskrit, our fellowship within the watertight world of neatly stacked journals and equations pale beside that final image, that final search for completion.

Seasons

In the fall of 1969, there were 500,000 American troops in Vietnam. The death of Ho Chi Minh caused only a brief interruption in the fifteen-year-old war. And I, beginning my senior year in college, was faced with the first real challenge to a life of privilege and ease. I was thrown into the national lottery for the draft. It was the first selective service lottery in the United States since 1942. World War II, of course, had been a "popular" war. My father had been constantly afraid of dying on the beaches of Italy or Sicily, but he had not hesitated to enlist when he came of age, nor had any of his friends. My friends, by contrast, did everything possible to escape military service. They were usually successful. Many got educational deferments simply for being in college. Louis, a quiet boy with a brooding intelligence, had dressed up as a Cherokee Indian for his physical examination, including war paint and feathers, and received a psychiatric release. Others moved to Canada and were sent money from home. But the new lottery seemed a great equalizer of classes and backgrounds. Everyone faced the same odds. Each birth date was to be assigned a number by a toss of the dice. Local draft boards would begin drafting at number 1 and work their way upward.

The drawing was made on December 1 at 8:00 P.M. Eastern Standard Time. Only the day before, on a Sunday, I had returned from Thanksgiving vacation and a grand dinner with my parents and brothers and cousins. After dinner, my mother, determined to be gay, placed a bossa nova album on the record player and made all of us dance with her barefoot in the living room. Now, a few evenings later, I sat anxiously with my roommates in our comfortable Ivy League dormitory room, listening to the radio. The scent of marijuana hung in the air. I imagined millions of other young men, short-order cooks in hamburger joints and gas station attendants trying to close for the night and other students in their rooms, all listening to their radios. Three hundred and sixty-six capsules were plucked from a cylindrical glass bowl in a government room in Washington, D.C. The first birthday chosen was September 14. I didn't know anyone born on that day, but I felt sorry for the poor devils. My birth date was chosen 280 draws later. I was never called. About a quarter of my classmates ended up in some kind of military service, that year or later.

Oddly, I remember that fall as intensely beautiful. Autumn

had never been a particularly engaging season in Tennessee, where I had grown up, but here, up along the East Coast, the air was so clear and transparent that you felt you might see to the curve of the earth. I recall often hearing an extraordinary concert from a maple tree outside my dormitory window. Hundreds of field finches had decided to roost in that tree for the season. The field finch is a small sparrow-sized bird with a delicate pointed tail. The bird does not twitter or chirp but instead gives out a continuous drawn-out song. When hundreds sing in unison, the sound is an unbroken chorus, with the effect on the hearing like that of a waterfall on the sight, a multitude of tiny droplets combining to make one sweeping flow. The birds stayed until the end of October, then one day were suddenly gone in their migration south.

The lottery disturbed me in many ways. I had lived a life of self-imposed blindness, not just the blindness that comes with financial good fortune and social entitlement. There was, of course, the real possibility of being sent to Vietnam and killed. But this outcome was so unimaginable that it never entered my consciousness. I had stood on the sidelines in naive disbelief as my classmates tried to batter down the front door to the Institute for Defense Analysis. I avoided the bonfires. When a young assistant professor sitting next to me at dinner one night lit a match to his draft card and invited us students to join him, I admired his boldness but didn't have a shred of understanding of what he had done. The lottery forced a vast, unwanted world on me, and the sensation was a painful gush of blood through the veins. Particularly distressing was the element of randomness, the uncertainty. I wanted to make decisions. I would go on to graduate school or I wouldn't. I would pursue a particular young woman or I wouldn't. I would leave my bicycle out in the courtyard at night or I would haul it down to the basement.

Science, for me, had been a source of certainty. I was a physics major, and physics reduced the world to its irreducible particles and forces. It is a banality to say that science holds a reductionist view of the world, and even a twenty-one-year-old knew that life wasn't so simple. But science, especially physics, provides a powerful illusion of simplicity and certainty. Textbooks on physics rarely offer any discussion of the history of the subject, with its wrong turns and prejudices and human passions. Instead, there are Laws. And the Laws seduce with their beauty and precision. Every action has an equal and opposite reaction. The gravitational force between two masses varies inversely with the square of the distance between them. Even Heisenberg's quantum Uncertainty Principle, which proclaimed that the future cannot be determined from the past, gave a definite mathematical formula for containing uncertainties, like a soundproof room built around someone who is screaming. More than its purity and grace, physics was Certainty. And Certainty, for reasons of my own temperament and perhaps also my middle-class upbringing, was my ally. Archimedes and Euclid had stood for Certainty. Lucretius had invoked the atomistic theory of the world in order to free humankind from the vagaries of the gods.

As a senior I was required to do an independent project, a

thesis. For some reason that I still cannot fathom, I chose to do an experimental thesis—that is, to build an apparatus for doing an experiment in physics. I had already shown myself completely incompetent in the laboratory. A gadget that I had constructed for my junior-year lab project caught fire because of faulty wiring. The oscilloscope, a standard tool for circuit design, a big metal box covered with knobs for adjusting voltages and currents, baffled me. On the other hand, I was good at theoretical calculations. I loved going from one equation to the next until arriving at the answer, as definite and unassailable as the area of a circle. I loved the cleanliness of pencils and paper. Why I didn't undertake a thesis in theoretical physics I do not know.

Perhaps it was my choice of thesis advisor, Professor Turgot. There was something about Professor Turgot that I found immensely appealing. He was a big bearlike man, forty-ish, beginning to bald, stoop-shouldered, whose shirttails always drooped down behind him. He was not at all the absentminded professor. He could fix me and all I was thinking with one eagle glance. When he lectured in the classroom, he addressed the blackboard rather than his students, as if he were having a private conversation with some mythical being living in the world he had created in equations and diagrams. I knew that this lecturing style was deficient, but it conveyed a lifelong fascination with his subject. I wondered whether I could contain my own passion for science, keep it from thinning out and dispersing for twenty more years, when I would reach the age of Professor Turgot.

Professor T was focused, but at the same time he was humble about the limitations of his knowledge. He sometimes confessed his professional blunders, an error in a calculation, a mispositioning of a target in the cyclotron. The rest of our teachers, almost without exception, projected the impression that they had gotten to where they were on a more or less laserlike trajectory. They had a magnificent self-confidence, which I am sure inspired many of their students. But even I, with my devotion to certainty, did not feel comfortable doing research with such a person. I knew that I made mistakes, and a thesis advisor who did so as well might allow me to graduate with my dignity. After class, Professor T, bulkily slumped against the wall and covered with chalk dust, would sometimes talk to me about his wife. Almost immediately, he began referring to her as Dorothy, so that when I finally met her, at dinner in the Turgots' small house, I felt as if I knew her. None of the other professors ever mentioned their spouses. I asked Professor T to be my thesis advisor. He grinned and said I would be doing an experimental thesis.

The laboratory where I began working was a huge cavern of a place, resembling a warehouse more than anything else. The space was filled with natural light, from skylights thirty feet overhead, as in an artist's studio. There was always an odd smell in the lab—not an unpleasant smell—of oil and dry ice. Canisters filled with liquid nitrogen sat on the concrete floor. When opened, these would emit a wonderful hissing noise as the liquid bubbled and evaporated and escaped in thick opaline clouds. Along three walls, stretching for a hundred feet,

were tabletops and workbenches, oscilloscopes, boxes of capacitors and resistors, odd pieces of metal, rubber tubing, Geiger
counters, notebooks with radioactive decay rates handwritten
in neat columns of figures. There were always a few novels by
Proust and Gide sitting casually on a lab table. Professor T's
wife, Dorothy, was a scholar in French literature. I like to
think that she sometimes visited the lab in the evening, to keep
her husband company when he worked there after hours.

In one corner of the lab, a shower faucet protruded inele-
gantly from the wall, in case someone accidentally came into
bodily contact with a radioactive substance and needed to strip
down immediately and wash off. The radiation shower I
noted with special interest, as I discovered that I had to con-
front radioactive atoms on a daily basis. My project was to
build a device capable of measuring the radioactive disintegra-
tion of excited states of neptunium. Neptunium, discovered in
1940, was the first chemical element produced artificially by
humankind. Since its atomic number, 93, was just beyond that
of uranium, 92, it was named for Neptune, the planet just
beyond Uranus. (Plutonium, at atomic number 94, was named
after Pluto.) The idea for my thesis, as it evolved in discussions
with Professor T, was that the excited neptunium would be
created by bombarding a uranium target in the cyclotron. The
disintegrating fragments of the neptunium nuclei, in flight
through my apparatus, would cause a gas to scintillate, and
these scintillations would be detected by several electronic
photomultiplier tubes. By carefully measuring the rate at
which neptunium nuclei fragmented, we could learn some-

thing about the forces struggling and churning within the atom.

As I stumbled along, writing up the specifications of various parts to be made in the machine shop and then respecifying when the parts didn't fit, I was helped by Dave, Professor T's assistant. Dave was indispensable. He thought the undergraduates were "bloody Communists," and he despised the bearded protest marches, but he was devoted to Professor T and his students, and he was the only person who could get the vacuum pump to behave. A vacuum pump, when working properly, starts out with a coarse, grating sound, like the chug of a locomotive, then graduates to a clicking whine, rising in pitch, and ends with a quiet, smooth hum when a good vacuum has been attained. When there is a leak in the system, the pump never progresses beyond the rough, grating chugs. On a number of occasions, I had to pump all the air out of my tangle of brass fittings and Mylar meshes, down to a billionth of an atmosphere. After applying epoxy and Glyptal to all the suspicious joints, we would turn on the vacuum pump. Dave understood that pump, as well as most things in the lab. His understanding went even further than that. I believe he was romantically involved with the woman who delivered small supplies to the lab. After her delivery each week, she would stand at an outside window and look in at him, sadly and longingly.

That winter Dave and I were often the only people in the lab, me puzzling over response curves of the photomultiplier tubes and him quietly fixing some gadget that had broken.

Occasionally I had to stop and walk over to an electric heater for warmth. Outside, the snow lay across the ground in a vast white silence. Then I would hear a squeaking and crunching, distant at first but gaining in volume, the sound of Professor Turgot's galoshes in the snow as he walked along the path from his office to the lab to check up on his charges.

My apparatus passed all of its preliminary tests, but I never did truly believe that the final experiment would work, and I don't think Professor Turgot did either. When it was time to insert the apparatus into the cyclotron in another building, I received a mysterious message that the cyclotron couldn't be scheduled until a few months after I'd graduated. "I'll write you about the results," Professor T kindly said to me and gave me high marks on my endless drawings of sideviews and topviews and calculations of solid angles and efficiencies. Professor Turgot never wrote and I never asked.

One spring afternoon, soon after Nixon had ordered the invasion of Cambodia, the department of physics had an extraordinary meeting. All the physics faculty and students crowded together into a small room to discuss our departmental response to the student riots taking place on campus. With neatly chalked equations still on the blackboard from some previous hour, faculty members got up one by one and delivered their views on the war. Most were strongly opposed but not all. There were brief and passionate speeches about the nature of democracy, the rights of governments, the purpose of education, moral responsibility. I could hardly recognize these people dressed up like our measured professors. The little room became a struggling upside-down box. I needed air. The discussion turned to a practical matter. What should the department do with its students who were cutting their classes? In the end, the faculty decided to exempt seniors from their final exams and, in some cases, their theses.

I reeled out of the room. To my dizzy and confused mind, randomness had finally won out. The world was a jumble of mistaken adventures, crossed wires, mirrors at odd angles. Certainty was a deception. And for me, at that moment in my life, there was either certainty or randomness, nothing in between.

I called Andrew, a roommate from freshman year and a quiet boy like myself. We walked to the lake a mile away and went sailing. It was early May. The breeze was so light that we finally lowered the sails and just drifted, half asleep in the hot thick air. We took off our shirts. Soon we were coasting near one of the shores, passing below willow trees that hung down into the boat and tickled our faces with their soft filigree leaves. Finally, a large branch got tangled in the mast and stopped our motion altogether, and we just lay there, enjoying the shade. I got up from my practically prone position and saw that our boat was surrounded by lilies, floating just next to the shore. A few had started to bloom, in luscious white flowers with a speck of purple at their centers. We lay there for hours.

And as we lay there, accidents happened all around us. A bird landed in a nearby tree for no reason and began singing, then flew off just as unexpectedly. Twigs snapped. Clouds changed shape. Grasses rustled with the movements of unseen animals. The earth wobbled imperceptibly on its axis, as bits of cosmic debris randomly bombarded it from space. One such piece of debris, billions of years in the past, had struck with unusual force and cocked the planet over, producing a tilt of twenty-three degrees, producing uneven heating as the earth orbited the sun, producing the seasons. A crumpled piece of paper slowly drifted past in the water, caught on a stick. Some writing on it had become smeared and illegible, perhaps a schedule of someone's appointments, or a note to a lover.

About the Author

ALAN LIGHTMAN was born in Memphis, Tennessee, in 1948 and was educated at Princeton and at the California Institute of Technology. He has written for *Granta, Harper's, The New Yorker,* and *The New York Review of Books.* His previous books include *Origins, Ancient Light, Great Ideas in Physics, Time for the Stars,* and two novels, *Einstein's Dreams* and *Good Benito.* He is Burchard Professor of Science and Writing, and senior lecturer in physics, at the Massachusetts Institute of Technology.